Enzyme Physics

Enzyme Physics

Mikhail V. Vol'kenshtein
Institute of Molecular Biology
Academy of Sciences of the USSR, Moscow

Translated from the Russian

Ⓟ Springer Science+Business Media, LLC 1969

Library of Congress Catalog Card Number 69-12546

The original Russian text was published by Nauka Press in
Moscow in 1967 as part of a series entitled Fizika Zhiznennykh
protsessov (Physics of Life Processes)

Михаил Владимирович Волькенштейн

ФИЗИКА ФЕРМЕНТОВ

FIZIKA FERMENTOV

ENZYME PHYSICS

ISBN 978-1-4899-2836-8   ISBN 978-1-4899-2834-4 (eBook)
DOI 10.1007/978-1-4899-2834-4

© 1969 Springer Science+Business Media New York
Originally published by Plenum Press, New York in 1969.

FOREWORD

This book treats a new, far-from-fully-developed area of molecular biophysics—enzyme physics. An attempt is made to survey this field, but primary consideration is given to three problems under investigation in the Polymer Structure Laboratory of the Institue of High-Molecular Compounds, Academy of Sciences of the USSR.

The first problem is the genetic coding of the biologically functional structure of proteins. Its solution is based on physical theories of hydrophobic interactions.

The second problem is the conformational properties of proteins as the factor governing enzyme activity. The most direct methods for experimental investigation of questions in this area are optical, principally those involving natural and magnetic rotation of the plane of polarization. A substantial portion of the book concerns optical activity; the Faraday effect is discussed in an appendix.

The third problem is the manifestation of the cooperative properties of enzymes in the kinetics of enzymatic reactions and the solution of complex kinetic problems. This problem is especially pressing in connection with research on allosteric enzymes, which are responsible for feedback in metabolic processes. An appendix describes a new method for solving kinetic problems, based on the theory of graphs.

This volume extends and details certain of the ideas expressed in my previous book, Molecules and Life: An Introduction to Molecular Biophysics, which was published in this series in 1965.

The writing of this book was directly connected with my participation in two winter schools on molecular biology in Dubna, both as a lecturer and as an auditor.

I wish to thank my colleagues and graduate students I. A. Bolotina, B. N. Gol'dshtein, D. S. Markovich, and Yu. A. Sharonov for their assistance in the preparation of this work. Special thanks is due to Academician A. E. Braunshtein for the valuable criticisms he offered after reading the manuscript.

M. Vol'kenshtein

Leningrad, May 1966

CONTENTS

Chapter 1

INTRODUCTION

Enzymes are proteins that serve as catalysts in biochemical processes. The catalytic function of proteins is of the greatest importance for life. In crude terms, the molecular plan for cellular and organismic structure laid down in the genome consists principally in a program for synthesis of definite enzymes at given stages of development. A single gene corresponds to a single protein chain. The existence of this program is also the basis of heredity. The catalytic activity of enzymes is responsible for the synthesis of all biologically functional compounds in the cell, for the storage of chemical energy (mainly in molecules of adenosine triphosphate, ATP), for the expenditure of this energy in biosynthetic processes, mechanochemical processes, and the active transport of molecules and ions, etc. As has recently been established, special enzymes, referred to as allosteric enzymes, regulate multistage biochemical processes through feedback. The cybernetic processes in the cell (information storage and transfer, regulation, and control) have a chemical character and are effected at the molecular level. This distinguishes the living cell or organism from any modern machine, in which such processes are carried out through mechanical, electrical, or magnetic phenomena. Since all chemical reactions in the cell involve the direct participation of enzymes, their significance in biology cannot be overemphasized. Enzymes carry out their function either in the form of free molecules in a homogeneous medium (an aqueous solution of inorganic ions and complex organic compounds) or as components of condensed supermolecular structures, which are also in contact with the aqueous medium. Examples of dissolved enzymes include the proteolytic enzymes in the digestive system, while examples of enzymes in condensed structures include the cytochromes in the mitochondria and myosin in muscle

1

fibers. There is at present no reason to suppose that enzymatic activity differs in these two cases, but we cannot exclude the possibility that such differences will be discovered during further studies of supermolecular structures. In any event, the mechano-chemical properties of the supermolecular structures forming cell and organoid membranes and contractile systems (muscle, etc.) are inseparably associated with the functions of the enzymes they contain. Enzymes differ from all other catalysts known to chemistry in their exceptionally efficient operation under mild conditions (in an aqueous medium at normal pressure and a low, physiological temperature). At the same time, most enzymes are very specific, acting selectively on definite compounds (substrates) or groups of compounds. Research on the structure and properties of enzymes is naturally an important and extensive branch of modern science and is called enzymology. This field is a component area of biochemistry and a great deal has been achieved in it. General theories have been devised to account for the mechanism and kinetics of enzymatic catalysis and advanced methods have been developed for studying this process. The detailed chemical reaction mechanisms of specific enzymes have been investigated in a number of cases. The molecular structure of many enzymes has been studied and the primary structure of ribonuclease, cytochrome c, chymotrypsin, and lysozyme has been completely deciphered. The three-dimensional structure of lysozyme and ribonuclease has been determined. Physicochemical hypotheses have been advanced to more or less successfully account for the special properties of enzymes as biological catalysts and regulators.

The development of biology since mid-century has led to a close and fruitful cooperation with both chemistry and physics in the solution of a number of extremely important problems. Application of the ideas and methods of modern physics to biology has proved to be very useful. We can now speak of enzyme physics as an important branch of molecular biophysics, inseparably associated with biochemistry and enzymology.

The problems of enzyme physics coincide with certain biochemical problems, but various aspects of research in these areas differ.

An enzyme is a high-molecular protein. Proceeding from the general premises of macromolecular physics, we must study the secondary, tertiary, and quaternary structure of the protein and the nature of the forces governing these structural levels and investigate the thermodynamics and kinetics of the function-related changes in structure. This can be accomplished by the methods of experimental and theoretical physics. Study of structure, thermodynamics, and kinetics constitutes the three main branches of molecular physics and thus of enzyme physics.

The direct action of an enzyme on a substrate is chemical in character and is determined by the specific properties of the various amino acid residues in the protein and by those of the prosthetic groups and coenzymes. The physical problem is to investigate the general properties of an enzyme as a dynamic macromolecular system.

The natural assumption that there are more or less universal physical mechanisms of enzymatic activity is a hypothesis whose validity has in no way been proved. The nature of enzymatic processes may perhaps be so diverse that general physical mechanisms do not exist, the situation reducing to a special type of chemistry for each individual case. Nevertheless, there is no other physical pathway open: one must seek general mechanisms in the processes studied and avoid assigning them extreme individualization.

If it is found in the course of such research that the hypothesis is invalid, this negative result will also be important for science. Use of the experimental techniques of macromolecular physics and of optical and spectroscopic methods has contributed greatly to our understanding of enzymes.

The kinetics of enzymatic reactions is a traditional and highly developed area of enzymology. However, theoretical-analysis methods based on new mathematical algorithms are very important in this field.

Experimental physics thus studies the structure of enzymes and the equilibrium and kinetic characteristics of the changes in these compounds. Theoretical physics considers the nature of en-

zymes as macromolecular systems and develops new mathemati-
cal-analysis methods applicable to the kinetics of enzymatic reactions and to quantum-mechanical investigations of specific chemical processes. Quantum mechanics is also necessary for interpreting the optical and spectral properties of molecules and ex-
plaining intra- and intermolecular interactions. Modern macromolecular physics rests on the techniques of classical thermodynamics, statistical physics, and kinetics. In the material that follows, we will deal principally with the macromolecular aspects of
enzyme physics, using quantum mechanics to interpret optical
properties and not considering the problems of quantum biochemistry.

As we mentioned above, utilization of the notions and methods of physics for solution of biological problems has proved fruitful in many cases. On the other hand, biology has stimulated the
development of a number of physical disciplines; thus, the theoretical and experimental study of optical activity has been greatly
advanced in connection with work in molecular biophysics.

The boundaries between biology, chemistry, and physics in
this area are extremely vague and the terminology used is accordingly somewhat arbitrary. What distinguishes molecular biophysics from biophysical chemistry? The same problems are studied
with the same research techniques in both cases. However, the
term "molecular biophysics" has some justification: if biophysics
and molecular physics exist, there is also a molecular biophysics.
However, this is not of great importance. What is important is
that the interaction of biology with chemistry and physics is extremely useful and promising. Research must be many-faceted.
As Lichtenberg [1] wrote: ". . . he who understands nothing but
chemistry does not sufficiently understand that."

Chapter 2

PROTEIN STRUCTURE

The biological properties of proteins, particularly their enzymatic activity, are governed by the special structure of protein molecules, which differs materially from the structure of any other compounds of natural or synthetic origin.

Proteins are macromolecular compounds consisting of copolymers (or, more precisely, copolycondensates) of 20 types of amino acids. The amino acid residues in proteins are very diverse and contain different functional groups. Table 1 gives the structural formulas of the amino acid residues, while Table 2 classifies them in accordance with a number of traits. The amino acid residues of the polypeptide chains of proteins are joined by peptide bonds, which have the form CO—NH— (with one exception: proline

is joined to adjacent residues by the $-\mathrm{CO-N}{\Big\langle}$ group).

Specific functionality inheres in the residues, but the properties of the peptide groups, which are identical for all the residues, are also very important. Their most significant characteristic in

this sense is their ability to form hydrogen bonds, $-\mathrm{OC}-\overset{|}{\mathrm{N}}-\mathrm{H}...$

$\mathrm{O}=\overset{|}{\mathrm{C}}-\mathrm{NH}-$ (the peptide group of proline does not contain hydrogen).
We can thus point out three basic factors governing protein structure: (1) proteins are macromolecular; (2) a protein chain consisting of n amino acid residues contains n − 1 peptide groups, which are capable of forming hydrogen bonds with one another, with water molecules, and with other compounds; and (3) the side-chain groups, or amino acid radicals, have very diverse structures and differ in physical and chemical functionality. Let us now consider the role of these factors.

5

Table 1. Amino Acid Residues Comprising Proteins

No.	Name	Symbol	Structural formula
1	Glycine	Gly	$-CO-CH_2$, with NH below
2	Alanine	Ala	$-CO-CH-CH_3$, with NH below
3	Valine	Val	$-CO-CH-CH\begin{smallmatrix}CH_3\\CH_3\end{smallmatrix}$, with NH below
4	Leucine	Leu	$-CO-CH-CH_2-CH\begin{smallmatrix}CH_3\\CH_3\end{smallmatrix}$, with NH below
5	Isoleucine	Ile	$-CO-CH-CH\begin{smallmatrix}CH_2-CH_3\\CH_3\end{smallmatrix}$, with NH below
6	Phenylalanine	Phe	$-CO-CH-CH_2-C\begin{smallmatrix}CH-CH\\CH=CH\end{smallmatrix}CH$, with NH below
7	Proline	Pro	$-CO-CH-CH_2$, $N-CH_2$ CH_2 (ring)
8	Trytophan	Trp	$-CO-CH-CH_2-C-C\begin{smallmatrix}H\\C\end{smallmatrix}CH$ ring with HC, C, CH, CH, N, H; NH below
9	Serine	Ser	$-CO-CH-CH_2OH$, with NH below
10	Threonine	Thr	$-CO-CH-CH\begin{smallmatrix}OH\\CH_3\end{smallmatrix}$, with NH below

Table 1 (continued)

No.	Name	Symbol	Structural formula
11	Methionine	Met	$-CO-CH-CH_2-CH_2-S-CH_3$ with NH below
12	Cystine	Cys	$-CO-CH-CH_2-S-S-CH_2-CH-NH$ with NH and CO below
12a	Cysteine	Cys-SH	$-CO-CH-CH_2-SH$ with NH below
13	Asparagine	Asp-N	$-CO-CH-CH_2-CO-NH_2$ with NH below
14	Aspartic acid	Asp	$-CO-CH-CH_2-CO-OH$ with NH below
15	Glutamine	Glu-N	$-CO-CH-CH_2-CH_2-CO-NH_2$ with NH below
16	Glutamic acid	Glu	$-CO-CH-CH_2-CH_2-CO-OH$ with NH below
17	Tyrosine	Tyr	$-CO-CH-CH_2-C$ (ring: $HC=CH$, $HC-CH$) $C-OH$, with NH below
18	Histidine	His	$-CO-CH-CH_2-C=CH$ with NH below, HN N ring with C H
19	Lysine	Lys	$-CO-CH-CH_2-CH_2-CH_2-CH_2-NH_2$ with NH below
20	Arginine	Arg	$-CO-CH-CH_2-CH_2-NH-C$ $\big\langle$ NH_2 / NH, with NH below

Table 2. Classification of Amino Acid Residues

I. By electrochemical properties

A. Neutral residues (14)

Ala, Asp-N, Val, Gly, Glu-N, Ile, Leu, Met, Pro, Ser, Thr, Trp, Phe, Cys

B. Acidic residues (3)

Asp, Glu, Tyr, (Cys-SH)

C. Basic residues (3)

Arg, His, Lys

II. By polarity

A. Polar (hydrophilic) residues (10)

Arg, Asp, Asp-N, Glu, Glu-N, His, Lys, Ser, Tyr (?), Thr, (Cys-SH)

B. Nonpolar (hydrophobic) residues (10)

Ala, Val, Gly, Ile, Leu, Met, Pro, Trp, Phe, Cys

III. By chemical structure

A. Aliphatic amino acid residues (15)

Ala, Arg, Asp, Asp-N, Val, Gly, Glu, Glu-N, Ile, Leu, Lys, Met, Ser, Thr, Cys, (Cys-SH)

Including:

a. Hydrocarbon residues (5)

Ala, Val, Gly, Ile, Leu

b. Residues containing a hydroxyl group (2)

Ser, Thr

c. Residues containing a carboxyl group (2)

Asp, Glu

d. Residues containing an amide group (2)

Asp-N, Glu-N

e. Residues containing an amino group (2)

Arg, Lys

f. Residues containing sulfur (2)

Met, Cys, (Cys-SH)

B. Amino acid residues containing π-electron rings (4)

His, Tyr, Trp, Phe

C. Imino acid residue (1)

Pro

The macromolecular structure provides for a potentially large number of conformations of the macromolecule as a whole. This results from the fact that the polypeptide chain contains simple C—CONH—C single bonds, around which internal rotation can occur. Let us assume that a polymer chain contains n single bonds and that the internal rotation about each of them is characterized by N energetically nonequivalent equilibrium positions, or rotary isomers [2-4]. The total number of conformations of the chain as a whole is N^n and is hence very large. A free polymer chain in solution should consequently be in the form of a randomly folded structure and not in the extended state. Actually, such a randomly folded structure corresponds to a lower free energy (higher entropy), since a folded structure with a distance between the ends of the chain far less than the chain length can be achieved by an infinitely larger number of conformations than the extended state. This is the case for most synthetic polymers. If a polymer is composed of inert groups containing only hydrocarbon radicals (e.g., polyethylene), no restrictions other than steric limitations and limitations resulting from the interaction with the solvent are imposed on the possible chain conformations. If the groups are functional, the shape of the randomly folded structure depends on their interaction. Restrictions are specifically imposed on protein conformations by intramolecular hydrogen bonds.

It is obvious that the disordered, fluctuating structure of a randomly folded protein contradicts the precisely determined biological functionality of the macromolecule. A protein can therefore have such a structure only in the denatured, biologically nonfunctional state. Macromolecules of synthetic polymers are similar to protein macromolecules in two respects: their macromolecularity and the large number of possible conformations in the absence of specific limitations. In this sense, synthetic-polymer macromolecules can serve as models of the extreme denatured states of proteins. The possible conformations are limited by the hydrogen bonds between the peptide groups (factor 2) both in proteins and in far simpler macromolecules, i.e., synthetic polyamino acids. These polypeptide chains are composed of residues of the same type. In some cases, polyamino acids and their esters occur in their ordered native forms in solution. In other words, they are stabilized in states with a definite conformation.

On the basis of x-ray diffraction data on the nature of peptide bonds, Pauling and Corey proposed a theoretical model of the polypeptide chain, the so-called α-helix, which is shown in Fig. 1 [5]. In the α-helix, the carbonyl of one peptide group is joined by a hydrogen bond to the imine radical of another peptide group.

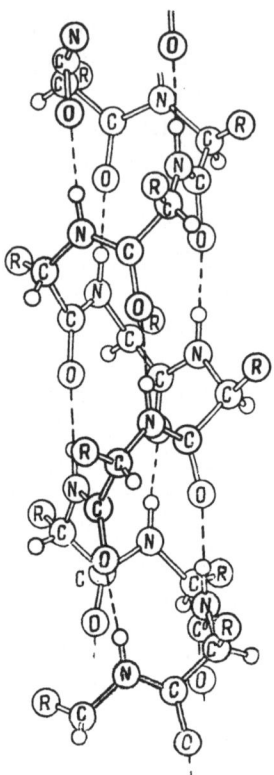

Fig. 1. Pauling–Corey α-helix.

These hydrogen bonds are directed along the axis of the helix and link the first peptide group to the fifth, the second to the sixth, etc. One turn of the helix contains 3.6 monomeric units, while the distance along the helix axis corresponding to one turn is 5.4 Å. The helix diameter is 10.1 Å [3, 6].

Utilizing hydrodynamic and optical methods, Doty demonstrated that a number of polyamino acids exist in α-helical form in solution. This is specifically true of poly-γ-benzyl-glutamate and polyglutamic acid. The α-helical conformation, fixed by hydrogen bonds, remains stable in certain solvents over a definite pH range at temperatures below the characteristic breakdown ("melting") temperature of the helix, at which the internal hydrogen bonds are broken and the macromolecule passes into a randomly folded structure. The helix–random coil transition is a cooperative process similar to the melting of a crystal [3, 4, 7, 8]. It is accompanied by a sharp decrease in viscosity, an abrupt change in the optical activity of the polypeptide, etc. It can be induced both by heating and by a change in solvent composition or medium pH (in the case of a polyelectrolyte). The statistical thermodynamic theory of equilibrium helix–random coil transitions has been worked out in detail (see [3, 4]), but the kinetics of this process still have not been sufficiently well investigated.

The existence of α-helical segments in native proteins has been proved by direct x-ray diffraction studies of myoglobin, hemoglobin, and lysozyme, and by a number of indirect procedures, the most important of which is spectropolarimetry (see pp. 1 and 2). The fraction of residues present in helical regions in proteins

Table 3. Degree of α-Helicity of Native Proteins (Average Data
Obtained by Various Spectropolarimetric Methods)

Native protein	Degree of α-helicity	Native protein	Degree of α-helicity
Paramyosin	0.96	Globin M.	0.47
Tropomyosin	0.87	Ovalbumin.	0.45
Light meromyosin,		Lysozyme	0.35
fraction I	0.87	Pepsin.	0.28
Myosin	0.60	Histone.	0.25
Heavy meromyosin	0.52	Ribonuclease	0.17
Insulin	0.51	Globin N.	0.12
Bovine serum albumin . .	0.50		

varies over a wide range. It can be zero, but, in contrast to polyamino acids, can never reach 100%. Table 3 presents data on the
α-helicity of certain proteins.

Synthetic polyamino acids can thus serve as good models for
studying the definite type of ordering inherent in native proteins.

Stabilized polypeptide-chain conformations other than the
α-helix are possible. In particular, the hydrogen bonds in certain
fibrillar proteins link peptide groups in adjacent chains and are
perpendicular to the chains. This is the polypeptide β-structure
and occurs in appropriate supermolecular structures. A "cross-
β-structure" is sometimes found in dissolved protein molecules
and certain polyamino acids; it differs from that described above
only in the fact that the transverse hydrogen bonds link different
segments of the same chain, which has an "accordion-like" arrangement (Fig. 2).

Let us now turn to the third factor, i.e., the diverse physicochemical functionality of the amino acid residues.

There can be different types of interactions among amino
acid radicals. Cys residues are linked to one another by disulfide
(S—S) bonds. This direct chemical interaction is specific for
these residues. Hydrogen bonds are formed between residues
containing OH and NH_2 groups, while dipole—dipole and inductive
interactions occur between polar residues. Ionizable acidic and
basic groups are joined by ionic (salt) bonds; these groups are
sources of ion—dipole and inductive interactions. The van der

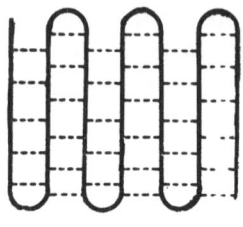

Fig. 2. Cross-β-structure
(diagram).

Waals dispersion forces between residues containing aromatic residues are especially strong. Finally, nonpolar residues participate in hydrophobic interactions, which play an especially important role in protein structure. The following chapter is devoted to such interactions. As a result of this complex interplay of forces, a protein macromolecule acquires a more or less compact globular structure in solution. The native form of an enzyme, which functions in solution rather than in a supermolecular system, is generally globular. The globule can contain both ordered (α-helical, cross-β-form, etc.) polypeptide-chain segments and disordered units having a certain conformational lability. The three-dimensional structure of a globular protein is not absolutely rigid and partial conformational rearrangements can take place. Figure 3 shows the structure of a myoglobin globule, as determined by x-ray diffraction analysis [9-11]. The helical and unordered segments can be seen. This protein has substantial helicity, reaching 75 % (see below, p. 28, for details).

Figure 4 is a diagram of the three-dimensional structure of an enzyme, lysozyme, which was the first enzyme structure to be determined [12]. The maximum possible degree of helicity in this case is 42%. The structure of lysozyme is more open than that of myoglobin.

Protein structure is described with the terminology introduced by Linderstrøm-Lang [13], who proposed a hierarchy of globular-protein structures. The primary structure is the sequence of amino acids in the polypeptide chain, the secondary structure is the conformation stabilized by the interpeptide hydrogen bonds (e.g., the α-helix or cross-β-structure; see Figs. 1 and 2), and the tertiary structure is the three-dimensional structure of the globule as a whole (Figs. 3 and 4). Finally, some protein macromolecules consist of two or more globules (subunits) that interact with one another. This is the quaternary structure. It is actually somewhat arbitrary to consider the secondary and tertiary structures separately, since any change in tertiary structure entails a change in secondary structure, and vice versa.

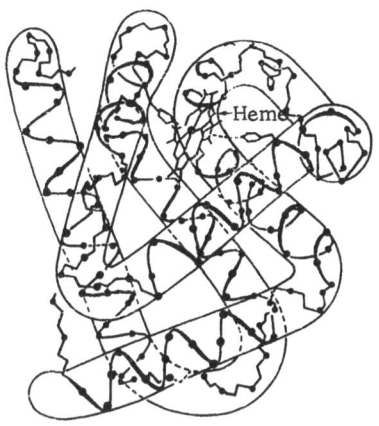

Fig. 3. Structure of myoglobin.

It can be seen that the three-dimensional structure of a protein is very specific and is governed by the three aforementioned factors (see p. 5). The role of the first factor can be shown by studying any synthetic polymer, while the role of the second can be elucidated by studying synthetic polyamino acids. It is particularly difficult to model the third factor, i.e., the diversity or polyfunctionality of amino acid residues. One is forced to model the individual types of interactions, by investigating the behavior in solution of different high- and low-molecular-weight compounds, electrolytes and nonelectrolytes, polar and nonpolar compounds, etc.

The physical theory of tertiary (globular) protein structure has encountered fundamental obstacles. Statistical thermodynamic analysis is applicable to an ensemble consisting of identical elements. The theory of the properties of synthetic macromolecules in solution is based on considering the monomeric units of the chain to be elements of this type. This so-called rotational isomeric model has been used to construct a theory that provides a good explanation for the properties of macromolecules in solution and for the elasticity of rubber-like polymer materials. The theory of helix—random coil transitions in synthetic polyamino acids deals with an ensemble of identical peptide bonds. The rotational—isomeric theory of macromolecules and the theory of helix—random coil transitions have successfully employed the mathematical apparatus of statistical mechanics, which was devised for investigating cooperative systems and rests on Ising's one-dimensional model [2-4]. The advances made in these theoretical investigations resulted from a refusal to individualize the elements of the statistical ensemble. We do not encounter such individualization for synthetic homopolymers. On the other hand, the three-dimensional structure of a protein depends to a substantial extent on the individual properties and arrangement of the 20 different types of

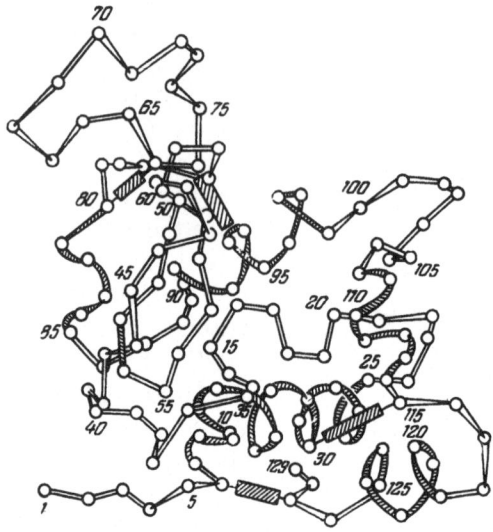

Fig. 4. Structure of lysozyme. The bonds in the α-
helical segments are shaded, while the disulfide bonds
are indicated by rectangles.

amino acid residues. There are at present no techniques for statistical analysis of this type of system. Instead of a strict theory,
we must therefore make use of various approximations, which enable us to interpret the principal factors responsible for the spatial organization of the protein molecule.

All that has been said above about the three-dimensional
structure of proteins pertains to the native state, in which proteins
exhibit their maximum biological functionality and hence their
maximum enzymatic activity. A denatured protein partially or
completely loses its functionality. The term "denaturation" refers
to those processes that cause a protein to lose its biological function as a result of disruption of its higher structural levels. The
peptide bonds in a denatured protein are the same as in the native
compound. In other words, the primary structure is retained during denaturation.

It is the primary structure of a protein, i.e., the sequence
of amino acid residues in the polypeptide chain, that is genetically
encoded. At the same time, the higher levels of protein structure
are directly responsible for biological functionality. One of the

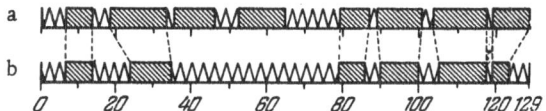

Fig. 5. Predicted (a) and observed (b) sequence of α-helical
segments (shaded rectangles) in lysozyme (after Guzzo).

most important and fundamental problems in molecular biology is
that of whether the primary structure of a protein predetermines
its three-dimensional structure. A positive answer to this question would mean that the three-dimensional structure is genetically encoded in the sequence of DNA nucleotides and that modern
molecular genetics thus rests on a sound footing. On the other
hand, a negative answer would mean that it is necessary to seek
additional special mechanisms for the encoding of higher structural levels and the genetic role of the nucleic acids would have
to be regarded as secondary. This dilemma can obviously be resolved by studying denaturation phenomena.

It has proved possible to demonstrate that denaturation is
reversible in a number of cases. Ribonuclease denatured without
rupture of its disulfide bonds [14], takaamylase [15], and other
proteins have been renatured. Anfinsen conducted an especially
thorough investigation of ribonuclease and showed that, even after
its disulfide bonds have been broken and reduced to free sulfhydryl
(S—H) groups, reoxidation leads to renaturation of the enzyme [16-
18]. Denaturation of alkaline phosphatase has been proved to be
reversible [19]. Similar phenomena are also known. Denaturation is theoretically reversible if there is no simultaneous break-
age of peptide bonds. Protein renaturation is possible in the absence of kinetic obstacles. Renaturation prevents aggregation of
the denatured molecules, formation of incorrect disulfide bonds,
etc.

On the other hand, even slight changes in primary structure,
such as replacement of one amino acid residue by another, lead to
substantial changes in biological functionality, i.e., to changes in
three-dimensional structure (see [3, 6]). A genic mutation in DNA
thus causes a change in the biological properties of the organism
and produces variation.

There are consequently solid grounds for assuming that the primary structure of a protein predetermines its higher structural levels. We know of no data contradicting this important assumption.

This gives rise to a clearly formulated problem of molecular physics: what three-dimensional structure corresponds to a given primary structure under equilibrium physiological conditions? The difficulty in fully resolving this question currently lies principally in the lack of detailed information on the three-dimensional structure of proteins. Sufficiently precise data have been obtained only for myoglobin, hemoglobin, and lysozyme. Nevertheless, important investigations that permit us to approach a solution have been conducted. Guzzo analyzed the primary and three-dimensional structure of myoglobin and hemoglobin, which had previously been studied by Kendrew and Perutz (see Fig. 3 and pp. 28 and 133), and established that the α-helical segments are interrupted by Pro, Asp, Glu, and His residues [20]. The fact that Pro cannot be incorporated into the α-helix was previously known, being obvious from the fact that proline is an imino rather than an amino acid residue. Guzzo attributed the behavior of Asp, Glu, and His to the charges they bear, which organize an oriented cluster of water molecules that hampers formation of an α-helix. Lys and Arg can also carry charges, but these are remote from the polypeptide backbone and do not create such obstacles. Knowing the primary structure of lysozyme, Guzzo predicted the distribution of α-helical segments along its chain. This prediction is in remarkably good agreement with recent data on the three-dimensional structure of lysozyme [12]. Figure 5, which is taken from Guzzo's article, compares the predicted and experimental results. On the basis of this type of data and current hypotheses regarding hydrophobic interactions, it is possible to devise a rather convincing model of the three-dimensional structure of a protein (see p. 27).

Blout previously gave a systematic description of the properties of polyamino acids [21]. Polyamino acids containing Val, Ile, Pro, Ser, Thr, and S-methylcysteine have no α-form. Hence it can be concluded that Val, Ile, Pro, Ser, and Thr interrupt the α-helix. Actually, Davies found that there was an inverse correlation between the content of these residues and the degree of protein α-helicity, as determined by spectropolarimetry [22]. Hav-

steen later obtained similar results for Pro, Ser, and Thr [23].
However, these data were obtained by an indirect technique and
are not in agreement (except for the data on Pro) with the results
of direct x-ray diffraction studies of myoglobin, hemoglobin, and
lysozyme.

Saroff attempted to establish the three-dimensional structure of ribonuclease theoretically, proceeding from the fact that
electrostatic interactions can occur between its charged groups
[24]. The degree of α-helicity of this enzyme is very low and may
even be zero. Saroff's model is close to an ellipsoid with dimensions of $(21-27) \times (25-31) \times (40-50)$ Å^3. The experimental data
vary, including $18 \times 30 \times 48$ Å^3, $30 \times 30 \times 38$ Å^3, and $31 \times 39 \times$
54 Å^3. Nevertheless, there is rough agreement. The principal
shortcoming of Saroff's work [24] is the fact that he ignored hydrophobic interactions.

The most general picture of three-dimensional protein structure can be obtained by considering hydrophobic interactions.

Chapter 3

HYDROPHOBIC INTERACTIONS AND PROTEIN STRUCTURE

In devising a theory to account for the three-dimensional structure of proteins, one must be guided by physical concepts of the interactions that stabilize the globule. The formation of ionic aggregates, which is considered by Saroff [24], cannot be regarded as the principal factor responsible for such stabilization and is not sufficient to account for it. Generally speaking, one cannot draw any serious conclusions regarding the three-dimensional structure of a protein without taking into account the actual conditions under which its native structure exists in the aqueous medium. A biologically functional protein exists in water and not in isolation. Water is a liquid having very special properties and the aqueous environment has a decisive effect on protein structure.

Water affects principally the hydrogen bonds in the globule. For stabilization of the secondary and tertiary structure, it is necessary that formation of hydrogen bonds of the $NH\cdots OC$ type be energetically more advantageous than formation of hydrogen bonds with the water molecules ($NH\cdots OH_2$ and $CO\cdots H_2O$). Klotz and Franzen [25, 26] established that formation of hydrogen-bonded methylacetamide dimers ($H_3C-CO-NH-CH_3$) is energetically advantageous in organic solvents but not in water. The heat of formation of the dimer is 4.0 kcal/mole in CCl_4, 1.6 kcal/mole in $CHCl_3$, and close to zero in water. Ptitsyn and Skvortsov estimated the energy of hydrogen-bond formation in polyglutamic acid and polylysine, making a theoretical analysis of the conditions for the helix—random coil transition in aqueous solution [27]. The bonding energy is small in these cases, amounting to about 0.2 kcal/mole of bonds. It can be assumed that other factors are more important

in structural stabilization. As long ago as 1944, Bresler and Talmud suggested that hydrophobic interactions constitute the main
factor responsible for development of the globular structure [28,
29]. Nonpolar amino acid radicals of the hydrocarbon type are
preferentially in contact with one another and not with water; on
the other hand, polar radicals interact with water. As a result, a
flexible macromolecule containing both types of radicals folds into a globule. The hydrophobic nonpolar radicals lie inside the
globule, while the hydrophilic polar radicals are at its surface.

This notion is based principally on the low solubility of nonpolar compounds, such as hydrocarbons, in water. On the other
hand, a number of data indicate that molecules containing both
polar and nonpolar groups are arranged in an aqueous medium in
such fashion that the polar groups are in contact with the water,
while the nonpolar groups are isolated from it. The classical experiments conducted by Langmuir, who studied monomolecular
layers of fatty acids on the surface of water, showed that the polar
carboxylic groups of the layer were immersed in the water, while
the nonpolar hydrocarbon radicals faced the air [30]. This phenomenon is responsible for the structure of the micelles in col-
loidal aqueous solutions of soap and other compounds; the hydrophobic groups are inside the micelles and the hydrophilic groups
are at the surface.

Kauzmann provided detailed substantiation of the concept of
hydrophobic interactions as the most important factor responsible
for formation of the globular structure of proteins [31]. He did
not limit himself to qualitative statements, but made a number of
quantitative comparisons and estimates.

Hydrophobic interactions have an unusual physical nature.
They result from the special properties of water, which differ
from those of other liquids. The thermodynamic disadvantage of
having contacts between hydrophobic groups and water molecules
results from the increased free energy. As has been shown experimentally, the free energy rises because of the decrease in entropy. Table 4 shows the change in free energy, enthalpy, and entropy during solution of hydrocarbons in water, as determined
from the formula

$$\Delta F = -RT (\ln K - \ln x) \qquad (1)$$

Table 4. Change in Thermodynamic Characteristics During
Solution of Hydrocarbons in Water at 25°C

Hydrocarbon	ΔF, kcal/mole	ΔH, kcal/mole	ΔS, cal/mole · deg
Methane, CH_4.	3.0	−2.6	−18
Ethane, C_2H_6	3.8	−2.0	−19
Propane, C_3H_8	5.0	−1.8	−23
Butane, C_4H_{10}.	5.8	−1.0	−23
Benzene, C_6H_6	4.3	0-0.6	−14
Toluene, $C_6H_5CH_3$.	5.0	0-0.6	−16

(where K is the equilibrium constant of the solution and x is the
molar concentration) and from the temperature dependence of K
and x.

Solution of nonpolar compounds is accompanied by a decrease in enthalpy, but this is overcome by a loss of entropy despite the positive entropy of mixing. The explanation of this effect must be sought in the structure of water. Water is a liquid
highly associated by hydrogen bonding. Water in the crystalline
state has a rather open structure, since the coordination number
in the lattice is low, i.e., four. Ice therefore has comparatively
low density, being less dense than water at 0°. The open, ice-like
structure is retained in liquid water: calculations have shown that
if this were not so and the water molecules were densely packed,
its density at 25°C would be 1.8 rather than 1.0 g/cm^3. The presence of a maximum (at 4°C) in the curve representing the density
of water as a function of temperature gives us grounds for considering water to be a system consisting of two structures: an
open, highly ordered, ice-like structure and a denser, disordered
structure. The former structure is characterized by lower enthalpy, entropy, density, and lability and a smaller coordination
number than the latter. Hydrophobic interactions can be treated
as a shift in equilibrium toward the more highly ordered and more
open structure, which is accompanied by a decrease in entropy.
This does not mean that "icebergs" are formed in proximity to the
hydrophobic groups: no ice crystals develop in this case.

There have been a number of attempts to devise a quantitative theory of water structure and hydrophobic interactions, of
which the most noteworthy are those of Nemethy and Scheraga [32,
33]. These authors calculated a partition function for water, con-

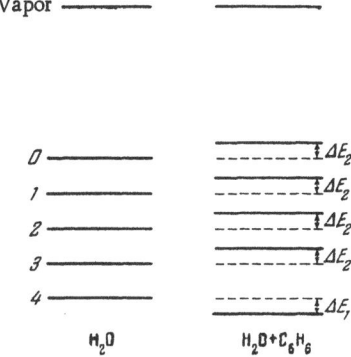

Fig. 6. Diagram of energy levels in water and aqueous hydrocarbon solutions (after Nemethy and Scheraga). The number of hydrogen bonds is indicated at the left.

sidering five possible states for the H_2O molecule: molecules having no hydrogen bonds and molecules forming one, two, three, and four such bonds. The latter state corresponds to ice. The coordination numbers of such molecules vary in reverse order, the highest number corresponding to the unbonded state, in which the molecules can be densely packed, and the lowest number (four) corresponding to the ice-like structure. Introduction of a hydrophobic group or molecule into water raises the coordination number of the molecules of the ice-like structure and thus reduces the energy of the molecules with four hydrogen bonds by a certain quantity ΔE_1. Conversely, the energies of all the other states increase (Fig. 6), presumably by the same amount ΔE_2. Utilizing these parameters, which have been found to have values $\Delta E_1 = -0.03$ and $\Delta E_2 = 0.31$ kcal/mole for aliphatic hydrocarbons and $\Delta E_1 = -0.16$ and $\Delta E_2 = 0.18$ kcal/mole for aromatic hydrocarbons, Nemethy and Scheraga obtained remarkable agreement between the calculated and measured thermodynamic constants of aqueous hydrocarbon solutions.

This theory cannot, however, be regarded as complete. It does not account for a number of properties, principally the water – ice transition at 0°C. These calculations must therefore be treated as rough estimates. It is important that they are based on intelligent molecular concepts.

Regardless of the theory employed, the experimental data enable us to consider the concept of hydrophobic interactions to be quite solidly grounded. It is obvious that the frequently employed term "hydrophobic bonds" is not precise. We are here dealing not with special bonds between nonpolar groups, but with nonspecific hydrophobic interactions resulting from changes in water structure. The free energy of the system

$$- CH_3 \ H_2O$$
$$H_2O \ H_3C -$$

is consequently higher than that of the system

$$\begin{array}{cc} -CH_3 & H_3C- \\ H_2O & H_2O. \end{array}$$

Like the interactions that occur in proteins, hydrophobic interactions can be modelled with the aid of water-soluble synthetic polymers. Such investigations have been undertaken only recently. Ptitsyn and his colleagues studied the properties of solutions of polymethacrylic acid (PMAA)

$$\left(\begin{array}{c} CH_3 \\ | \\ -CH_2-C- \\ | \\ COOH \end{array} \right)_n .$$

In contrast to polyacrylic acid (PAA)

$$\left(\begin{array}{c} H \\ | \\ -CH_2-C- \\ | \\ COOH \end{array} \right)_n ,$$

PMAA exhibits a number of peculiarities attributable to the hydrophobic interactions of macromolecules containing methyl groups [34]. In particular, the solubility of PMAA decreases as the temperature rises, like the solubility of hydrocarbons.

It can be assumed that hydrophobic interactions are the most important factor responsible for formation of protein globules: the hydrophilic polar residues are arrayed at the globule surface, while the hydrophobic nonpolar residues are inside the globule. On the basis of this hypothesis, we can ignore the diversity of the amino acid residues and separate them into only two groups: polar and nonpolar (see Table 2). We thus need consider only two rather than twenty types of elements in the ensemble. Let us see what results are yielded by this crude and simplified picture.

Fisher convincingly showed that such an approach enables us to understand the general characteristics of three-dimensional protein structure [35]. Assuming that all the amino acid residues have the same volume, we conclude that the outer layer of the globule is a monomolecular layer consisting of polar residues and having the constant thickness d. In the case of a spherical globule, we find the volume of this layer to be

$$V_e = \frac{4\pi}{3} [r^3 - (r-d)^3],$$ (2)

where r is the globule radius. The internal volume of the globule, which is filled with nonpolar residues, is

$$V_i = \frac{4\pi}{3} (r-d)^3,$$ (3)

and the total globule volume

$$V_t = V_e + V_i.$$ (4)

The ratio of the volumes occupied by the polar and nonpolar residues, i.e., the ratio of the number of polar residues to the number of nonpolar residues, is

$$p = \frac{V_e}{V_i},$$ (5)

and, for a sphere,

$$p_s = \frac{r^3}{(r-d)^3} - 1.$$ (6)

A protein can be characterized by easily determined quantities: the relative polarity p and the total number of residues (and hence the total volume V_t). These quantities are related. It follows from Equations (4) and (5) that

$$V_e = V_t \frac{p}{p+1}.$$ (7)

Moreover,

$$V_e = Ad,$$ (8)

where A is the surface area of the nonpolar core of the globule. Fisher assumes d to be 4 Å. We find from Equations (7) and (8) that

$$p = \frac{A}{\frac{1}{4}V_t - A} \quad (A \text{ in } Å^2).$$ (9)

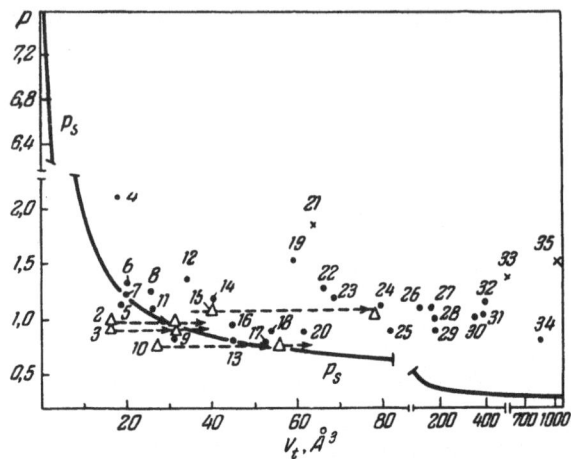

Fig. 7. Values of p (calculated from composition data) as function of. V_t for different proteins (after Fisher). The triangles represent aggregated proteins and the crosses represent proteins that have been found to have an elongated rodlike structure. 1) Salmine; 2) insulin; 3) insulin; 4) ribonuclease; 5) lysozyme; 6) myoglobin (whale); 7) myoglobin (horse); 8) papain; 9) chymotrypsinogen; 10) structural protein; 11) corticotropin; 12) ovomucoid; 13) pepsin; 14) carboxypeptidase; 15) prothrombin; 16) β-lactoglobulin; 17) pepsinogen; 18) ovalbumin; 19) edestin; 20) α-amylase; 21) tropomyosin; 22) plasma albumin (human); 23) plasma albumin (bovine); 24) avidin; 25) hemoglobin (horse); 26) conalbumin; 27) globulin (human); 28) aldolase; 29) thiophosphate dehydrogenase; 30) leucine aminopeptidase; 31) phosphorylase; 32) glutamate dehydrogenase ; 33) fibrinogen; 34) β-galactosidase; 35) myosin.

The smaller the volume V_t (i.e., the molecular weight of the protein), the greater should be its relative polarity p: there is no room for nonpolar groups to be arrayed in a small molecule. In the limiting case, we obtain $p_s \rightarrow \infty$ for a sphere with radius r = d (see [6]). Figure 7 shows the theoretical curve for a sphere and the experimental points for a number of proteins. The observed points are grouped near the curve but do not fall on it in most cases. A globule can be spherical only at $p = p_s$. For most proteins $p > p_s$, and the globule has a less symmetric form with a larger surface-to-volume ratio. The majority of the experimental points therefore lie above the theoretical curve. The molecule has a greatly elongated shape when $p \gg p_s$: the protein is fibrous rather than globular. This is true for tropomyosin and fibrinogen.

Conversely, when $p < p_s$, the polar residues do not cover the non-polar residues even with a spherical form. Nonpolar residues also occur at the globule surface and hydrophobic interactions lead to aggregation of the globules, i.e., to appearance of quaternary structure. Cases of this type are indicated in Fig. 7 by triangles; aggregation shifts these points to the right. Proteins can thus exist in aqueous solution only if Equation (9) is satisfied. It can be seen that very simple considerations based on the concept of hydrophobic interactions enable us to make a rough evaluation of the shape of a protein macromolecule solely on the basis of its . amino acid composition; we must know p and V_t (see also [317]). Fisher later considered the conditions for hydration of protein molecules [36]. He suggested that the protein surface is coated with a monomolecular water layer, in which the H_2O molecules are uniformly distributed. The number of grams of water per gram of protein is calculated in the following manner: a water molecule occupies a volume of 33.3 $Å^3$ and the equivalent surface area of the protein is 10.3 $Å^2$. The number of grams of water per $Å^2$ of protein surface, assuming a monomolecular layer, is

$$g = \frac{18 \text{ g/mole}}{10.3 \text{ Å}^2 \cdot 6.02 \cdot 10^{23} \text{ mole}^{-1}} = 2.9 \cdot 10^{-24} \text{ g per } 1 Å^2. \tag{10}$$

The hydrated surface area of the protein equals A. It follows from Equation (8) that

$$A = \frac{V_e}{d} = \frac{V_e}{4} Å^2.$$

The number of grams of water per gram of protein is

$$H_t = gA = \frac{gV_e}{4} = 0.725 \cdot 10^{-24} V_e = 0.725 \cdot 10^{-24} V_t \frac{p}{p+1}. \tag{11}$$

Table 5 shows the molecular weights, relative polarities, and values of H_t calculated from Equation (11) for 34 proteins.

The value of H_t is almost constant and averages 0.28 for proteins with molecular weights between 12,500 and 320,000. Given our premises, this means that the surface area of a protein macromolecule increases in proportion to its volume and that the A/V_t ratio is a constant independent of p (p ranges from 0.7 to 1.7 in Table 5). Comparison of the data in the table with the preceding calculations shows that p/p_s increases almost linearly as V_t (i.e.,

Table 5. Hydration of Proteins

Protein	Mol. wt.	p	$p/(p+1)$	$H_t, \dfrac{\text{g water}}{\text{g protein}}$
1. Cytochrome c.	12,500	1.601	0.62	0.316
2. Ribonuclease	12,700	1.709	0.63	0.330
3. Lysozyme	14,000	0.995	0.50	0.249
4. α-Lactalbumin.	15,500	1.105	0.52	0.285
5. Bovine myoglobin (adult)	17,500	1.193	0.55	0.292
6. Bovine myoglobin (fetal)	17,500	1.156	0.54	0.287
7. Papain	20,304	1.144	0.53	0.282
8. Chymotrypsinogen A (bovine)	25,100	0.852	0.46	0.239
9. Tryptophan synthetase . .	29,500	0.724	0.42	0.221
10. Carbonicanhydrase I (bovine)	31,000	1.260	0.56	0.295
11. Carbonicanhydrase II (bovine)	31,000	1.316	0.57	0.300
12. Insulin (bovine).	34,500	0.969	0.49	0.260
13. Carboxypeptidase B (bovine)	34,000	1.186	0.54	0.295
14. Carboxypeptidase B (hog)	34,000	1.139	0.53	0.289
15. Carboxypeptidase	34,300	1.212	0.55	0.297
16. Pepsin (bovine).	35,000	0.872	0.46	0.253
17. β-Lactoglobulin (bovine)	37,700	0.960	0.49	0.266
18. Ovalbumin (chicken) . . .	46,000	0.919	0.48	0.259
19. α-Amylase (B. subtilis). .	48,700	1.280	0.56	0.296
20. Inorganic pyrophosphatase	63,000	1.019	0.51	0.266
21. Plasma albumin (rabbit) .	65,000	1.206	0.55	0.290
22. Ovidin (egg yolk)	66,000	1.412	0.58	0.309
23. Enolase (yeast)	67,200	1.009	0.50	0.265
24. Hemoglobin (human) . . .	68,000	0.861	0.46	0.250
25. ATP-creatine trans-phosphorylase (rabbit). .	81,000	1.208	0.55	0.289
26. Lactic dehydrogenase (hog)	130,000	0.966	0.49	0.259
27. γ-Globulin (human). . . .	160,000	1.169	0.54	0.288
28. Glycogen phosphorylase (human)	242,000	1.072	0.52	0.275
29. Glycogen phosphorylase (rabbit).	242,000	1.085	0.52	0.278
30. Catalase (bovine)	250,000	1.171	0.54	0.284
31. Leucine aminopeptidase .	300,000	0.964	0.49	0.259
32. Phycocyanin (P. tenera) .	300,000	0.991	0.50	0.266
33. Heavy meromyosin (rabbit)	320,000	1.260	0.56	0.298
34. α-Lipovitellin (egg yolk)	320,000	1.094	0.52	0.276

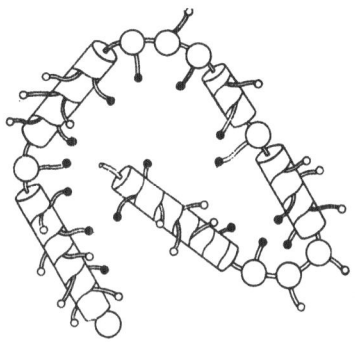

Fig. 8. Three-dimensional protein structure (after Guzzo). The black dots represent hydrophobic groups, the circles represent hydrophilic groups, the cylinders represent α-helices, and the spheres represent residues which do not form helices.

the molecular weight) rises. The larger the volume of a protein macromolecule, the more greatly its shape differs from spherical. It must be kept in mind that the volume V_t in Equation (11) is calculated from the molecular weight of the protein, assuming that the specific volumes (per unit molecular weight) of all the amino acid residues are roughly the same. It is actually found that the dependence of H_t on $p/(p + 1)$ is almost linear, which proves the validity of our hypothesis.

On the other hand, there are also very large proteins of spherical shape. The data in Table 5 are not exhaustive in this sense.

Nevertheless, the results of Fisher's work can be regarded as strong arguments in favor of the theory that hydrophobic interactions play a decisive role. Two factors are important here. First of all, it has been shown that consideration of two rather than twenty types of amino acids enables us to draw valuable inferences regarding protein structure. This to some extent eliminates the difficulties pointed out above and gives rise to the biologically important deductions made in the next chapter. Secondly, it follows from the agreement between the theoretical concepts and experimental results that the water surrounding a protein (or a nucleic acid) is not structured, i.e., does not have reduced entropy. The hydrophobic residues are concealed within the globule precisely because such structuring is thermodynamically disadvantageous. This refutes numerous works on the presence of structured water in biological systems and its special role in energy-migration processes, etc. (see [37]).

The true situation is of course considerably more complex. Calculation of the influence of hydrophobic interactions by Fisher's method is suitable only for rough estimates, since both the diversity of polar and nonpolar residues and the sequence of these resi-

Table 6. Substitution of "Internal" Amino Acid Residues of Hemoglobin and Myoglobin

Residue designation, after Kendrew and Perutz	Residues observed in different species	Residue designation, after Kendrew and Perutz	Residues observed in different species
A18	Val, Ile, Leu	E15	Val, Leu, Phe
A11	Ala, Val, Leu	E18	Gly, Ala, Ile
A12	Try, Phe	E19	Val, Leu, Ile
A15	Val, Leu, Ile	F1	Leu, Ile, Phe, Tyr (?)
B6	Gly	FG5	Val, Ile
B9	Ala, Ile, Ser, Thr	G5	Phe, Leu
B10	Leu, Ile	G8	Val, Leu, Ile
B13	Met, Leu, Phe	G11	Ala, Val, Cys-SH
B14	Leu, Phe	G12	Leu, Ile
C4	Thr	G16	Leu, Val, Ser
CD1	Phe	H8	Leu, Phe, Met, Trp, Tyr (?)
CD4	Phe, Trp		
D5	Val, Leu, Ile, Met	H11	Ala, Val, Phe
E4	Val, Leu, Phe	H12	Val, Leu, Phe
E8	Gly, Ala	H15	Val, Phe
E11	Val, Ile	H19	Leu, Ile, Met
E12	Ala, Leu, Ile	H23	Tyr (?)

dues in the polypeptide chain are ignored. It is obvious that non-polar residues can also occur at the globule surface if they adjoin hydrophilic polar groups. On the other hand, the globule core must be predominantly nonpolar, since this provides for the hydrophobic interactions that stabilize the compact structure. The model of three-dimensional protein structure proposed by Guzzo [20] takes this factor into account. Figure 8 is a diagram of his model. The cylindrical segments are α-helices, while the spherical segments contain Pro, Asp, and His. The hydrophobic groups, which are represented by black dots, face into the globule.

A detailed study of the structure of myoglobin and hemoglobin [38] has confirmed the important role of hydrophobic interactions in organizing the globule core. These proteins contain 33 "internal" residues not in contact with the ambient water. With three exceptions, the internal residues are always nonpolar in the most diverse species of vertebrates (primates, horses, hogs, rab-

bits, llamas, carp, and lampreys for hemoglobin, and man and
sperm whales for myoglobin). This is illustrated by Table 6.

The polar substitutes are underlined in the table. A question mark is used for tyrosine, which the authors consider to be
hydrophobic. Actually, Tyr exhibits both polar and nonpolar properties. Analysis of the structure shows that these 33 residues
are located on the side of the α-helix facing into the globule. On
the other hand, the residues at the globule surface do not obey the
simple rule pointed out above: polar residues can be replaced by
nonpolar residues, and vice versa.

The existence of a hydrophobic core in the globule has also
been confirmed by investigations of the denaturation of proteins by
slightly polar organic solvents [39]. In such cases, the nonpolar
groups of the protein are in contact with the nonpolar groups of
the solvent and the "hydrophobic structure" breaks down. The
strong denaturative effect of urea is attributable to this phenomenon. On the other hand, water molecules can be incorporated into the intraglobular space and even into the interglobular inter-
stices (when a quaternary structure is present). The rule of
thumb described above may not be observed in such cases.

In concluding this chapter, it should be emphasized that a
protein globule is a compact structure that differs materially
from a randomly folded coil. The amino acid residues within the
globule are in contact with one another. The globule can resemble
a drop of liquid in this sense (see p. 43).

Chapter 4

GENETIC CODING
OF PROTEIN STRUCTURE

A gene, or cistron, is responsible for the synthesis of a protein chain; the sequence of amino acid residues in the protein is governed by the sequence of DNA codons (nucleotide triplets) and corresponding mRNA codons [2].

Let us reiterate the main conclusions drawn from the material presented in the preceding chapters.

1. The higher levels of protein structure, i.e., the three-dimensional arrangement of a protein, are evidently predetermined by its primary structure — the sequence of amino acid residues in the polypeptide chain. This means that the three-dimensional structure of a protein is genetically coded.

2. Hydrophobic interactions are important for the three-dimensional structure of a protein, which, in first approximation, is governed by the relative polarity p.

Several interesting consequences follow logically from this. If the value of p is so important, it can be assumed that replacement of a polar residue by a polar residue or of a nonpolar residue by a nonpolar residue should have less effect on the structure and biological functionality of a protein than replacement of a polar residue by a nonpolar residue, or vice versa. It is therefore to be expected that related proteins from different species, differing in amino acid composition, should have similar sequences of polar (p) and nonpolar (n) residues. We have already seen that this hypothesis is valid for the n-residues within the globules of myoglobin and hemoglobin. Margoliash called attention to the consistent arrangement of the p-residues (basic) and n-residues in the

Table 7. Primary Structure of "Active Center" of Hemoglobins
and Myoglobin

Hemoglobin	Residue sequence	Sequence of p- and n-residues
Human, α-chain........	Lys—Gly—His—Gly—Lys—Lys	p n p n p p
Human, β-chain........	Lys—Ala—His—Gly—Lys—Lys	p n p n p p
Bovine, α-chain........	Lys—Gly—His—Gly—Glu—Lys	p n p n p p
Llama, β-chain........	Lys—Gly—His—Gly—Thr—Lys	p n p n p p
Sheep, α-chain.........	Lys—Gly—His—Gly—Glu—Lys	p n p n p p
Orangutan	Lys—Asp—His—Gly—Lys—Lys	p p p n p p
Carp, β-chain	Ala—Ala—His—Gly—Arg	n n p n p
Burbot..............	Arg—Trp—His—Ala—Glu—Arg	p n p n p p
Leghemoglobin.........	Thr—Gly—His—Ala—Glu—Lys	p n p n p p
Myoglobin...........	Lys—Lys—His—Gly—Val—Thr	p p p n n p

important respiratory enzyme cytochrome c in five species of vertebrates (human, horse, hog, rabbit, and chick) [40]. In this connection, he wrote: ". . . whether this evolutionary consistency results from the pressure of selection on the structural characteristics of cytochrome or, albeit only partially, from genetic variability. . .is a question that cannot at present be answered." It was subsequently shown that such an answer does indeed exist.

The sequence of residues near the "active center" of the hemoglobins of different species obeys the same rule [41]. The term "active center" refers in this case to the group of residues surrounding the His directly attached to the nonprotein prosthetic group (heme). Table 7 presents data illustrating this phenomenon [42]. The table also includes proteins related to hemoglobin but differing from it functionally (leghemoglobin and myoglobin). Despite the great diversity of species, from fish to human, the sequence of the p- and n-residues breaks down in only a few cases. The table shows 22 residue substitutions in comparison with the β-chain of human hemoglobin. Of these, 18 (81.8%) are of the p → p (45.5%) and n → n (36.3%) types, and only 4 are of the p → n and n → p types (9.1% each). Two of the four "incorrect" substitutions are in myoglobin.

In natural human-hemoglobin mutants, substitution of amino acid residues usually occurs with no change in polarity [41]. Data

Table 8. Mutational Substitutions in Human Hemoglobin

Type of hemoglobin	Chain	Residue number	Substitution	Substitution type
HbI	α	16	Lys → Asp	p → p
HbG (Honolulu)	α	30	Glu → Glu-N	p → p
Hb (Norfolk).	α	57	Gly → Asp	n → p
HbM (Boston)	α	58	His → Tyr	p → p
HbG (Philadelphia) . . .	α	68	Asp-N → Lys	p → p
HbO (Indonesia)	α	116	Glu → Lys	p → p
HbS	β	6	Glu → Val	p → n
HbC	β	6	Glu → Lys	p → p
HbG (San Jose)	β	7	Glu → Gly	p → n
HbE	β	26	Glu → Lys	p → p
HbM (Saskatoon).	β	63	His → Tyr	p → p
HbM (Zurich).	β	63	His → Arg	p → p
HbM (Milwaukee)	β	67	Val → Glu	n → p
HbD$_\beta$ (Punjab)	β	121	Glu → Glu-N	p → p
HbO (Arabia)	β	121	Glu → Lys	p → p

to illustrate this are given in Table 8 [43, 44]. Eleven of the fifteen substitutions, i.e., 74%, follow this rule.

Comparison of the amino acid sequences of different Bence-Jones proteins, which are analogous to light antibody chains, leads to similar conclusions. Many such examples could be given. There are also examples of a different type: we might specifically point out the insulins of different mammalian species, where the number of "incorrect" substitutions exceeds the number of "correct" substitutions. However, insulin is a small specific polypeptide and substitutions in it take place within a short amino acid sequence held in the folded state by a disulfide bridge. The rule under consideration therefore may not be observed in this case.

There are obviously two alternatives: either mutational substitutions are equally probable for all amino acid residues and a substantial majority of the "incorrect" mutations are lethal, so that the substitutions in question are not observed, or the properties of the genetic code are such that "incorrect" substitutions are less probable than "correct" ones. In order to decide which of these alternatives actually holds, we must consider the manner in which polar and nonpolar amino acid residues are coded.

x \ y	A	C	G	U	z
A	I Lys 1	V Thr 17	IX Arg 32	XIII Ile 16	A
	II Asp-N 2	VI Thr 18	X Ser 28	XIV Ile 12	C
	III Lys 3	VII Thr 19	XI Arg 24	XV Met 8	G
	IV Asp-N 4	VIII Thr 20	XII Ser 20	XVI Ile 4	U
C	XVII Glu-N 5	XXI Pro 21	XXV Arg 31	XXIX Leu 15	A
	XVIII His 6	XXII Pro 22	XXVI Arg 27	XXX Leu 11	C
	XIX Glu-N 7	XXIII Pro 23	XXVII Arg 23	XXXI Leu 7	G
	XX His 8	XXIV Pro 24	XXVIII Arg 19	XXXII Leu 3	U
G	XXXIII Glu 9	XXXVII Ala 25	XLI Gly 30	XLV Val 14	A
	XXXIV Asp 10	XXXVIII Ala 26	XLII Gly 26	XLVI Val 10	C
	XXXV Glu 11	XXXIX Ala 27	XLIII Gly 22	XLVII Val 6	G
	XXXVI Asp 12	XL Ala 28	XLIV Gly 18	XLVIII Val 2	U
U	XLIX Ochre 13	LIII Ser 29	VIII Meaningless? 29	IV Leu 13	A
	L Tyr 14	LIV Ser 30	VII Cys 25	III Phe 9	C
	Amber 15	I Ser 31	VI Trp 21	II Leu 5	G
	Tyr 16	II Ser 32	V Cys 17	I Phe 1	U

Fig. 9. Coding pattern. The p residues are shaded. Identical Roman numerals indicate "parallel" codons, while identical Arabic numerals represent additional "antiparallel" codons.

Even before a complete code key had been established, when the composition but not the nucleotide sequence of the codons was known, statistical analysis showed a pronounced difference in the coding of polar and nonpolar residues [41, 45]. It was found that, for the most part, polar residues are coded by codons containing adenine and cytosine, while nonpolar residues are coded by codons containing guanine and uracil. The ratio $\alpha = (A + C)/(G + U)$ is 1.86 for p-residues and 0.50 for n-residues, while the average for all residues is 1.00.

Nirenberg and his colleagues subsequently established a complete key to the genetic code. The codons for all amino acid residues and their nucleotide sequences are now known [46-48], which has permitted substantially more complete analysis of the problem [49].

Figure 9 shows the coding pattern. A table with the same arrangement of the first, second, and third nucleotides appeared in Pelc's article [50]. However, Pelc used the nucleotides in the sequence U, C, A, and G and paid no attention to the difference in the coding of p- and n-residues, considering only the metabolic origin of the amino acids. It is obvious that coding is not associated with intracellular amino acid synthesis. The manner in which the amino acids interact during protein biosynthesis is independent of their origin. In the key shown in Fig. 9, the nucleotides are arrayed in the sequence A, C, G, and U (corresponding to their alphabetical order). The difference in the coding of p- and n-residues is far more apparent in this case. It can be seen that the p- and n-residues are formed over continuous areas. Ochre and amber are arbitrary designations for the "meaningless" codons UAA and UAG, which signal termination of the protein chain.

Tyrosine is considered to be a polar residue, but it can exhibit hydrophobic properties and be located inside the globule (see p. 29). In the unionized state, the hydroxyl group of tyrosine forms a hydrogen bond with a carbonyl group, which can reinforce the structure of the globule core. On the other hand, Tyr at the globule surface forms a hydrogen bond with a water molecule. Phenol, a Tyr analog, is soluble in water. It must therefore be assumed that Tyr is a residue with intermediate properties and, for greater simplicity, we will treat it as polar.

It is extremely important that the three nucleotides in a codon are not equally functional. In a number of cases, the third nucleotide can be replaced without changing the amino acid encoded, while replacement of the first or second nucleotide does alter the acid encoded in the overwhelming majority of instances [49, 51]. If we designate the codon as xyz, x can be called the prefix, y the root, and z the ending [52]. Figure 9 shows that replacement of the ending z never causes a change in the polarity of the amino acid residue when x and y are left unchanged. In an $xyz_1 \rightarrow xyz_2$ substitution (where z_1 and z_2 are different nucleotides), the residue remains unchanged, a p-residue is replaced by a p-residue, or an n-residue is replaced by an n-residue. In dealing with this aspect of the problem, we can therefore limit ourselves to considering the 16 xy doublets rather than the 64 xyz triplets. These doublets form the key to the amino acid code shown in Fig. 10.

x \ y	A	C	G	U
A	Lys, Asp-N	Thr	Arg, Ser	Ile, Met
C	Glu-N, His	Pro	Arg	Leu
G	Glu, Asp	Ala	Gly	Val
U	Tyr	Ser	Cys, Trp	Leu, Phe

Fig. 10. Simplified coding pattern.

The difference in the coding of p- and n-residues can be seen in the fact that the residue is polar when y = A and nonpolar when y = U. The meaning of the statistical regularity pointed out above now becomes clear.

In most cases, single nucleotide substitutions in the codon, i.e., spot mutations, correspond to p → p or n → n substitutions, but not to p → n substitutions. The total number of possible substitutions is 64 · 3 · 3 = 576. When the ending z is replaced, seven substitutions cause conversion of a nonmeaningful (n) triplet to a meaningful (m) triplet, seven cause an m → n conversion, and two cause an n → n conversion. The remaining 176 substitutions have no effect on the polarity of the amino acid residue. When the prefix x is replaced, nine substitutions lead to an n → m conversion and nine to an m → n conversion. Of the remaining 174 substitutions, 114 do not alter the polarity of the residue and 60 change it. When the root y is replaced, seven substitutions lead to an m → n conversion, seven to an n → m conversion, and two to an n → n conversion. Of the remaining 176 substitutions, 74 do not alter the polarity of the residue and 102 change it. On the whole, 364 of the 526 substitutions that do not involve meaningless codons, i.e., more than two-thirds, retain the same residue polarity. In this sense, the genetic code is rather noiseproof. The most important role in the code is played by the root y, mutation of which produces the greatest chance of a p → n substitution.

If all residue substitutions were equally probable, the ratio of "correct" to "incorrect" substitutions (disregarding substitutions involving meaningless codons) would be 0.9 rather than 2.24. Actually, there are 20 amino acid residues, of which 10 are polar and 10 are nonpolar. Each polar residue can be replaced by nine p-residues and ten n-residues, while each nonpolar residue can be replaced by nine n-residues and ten p-residues.

With a nonfunctional ending z, other relationships are also possible in this sense. If the p- and n-residues in the table in

Fig. 10 were arranged in checkerboard fashion, the ratio of "correct" to "incorrect" substitutions would be 1.25 : 1. On the other hand, if the p-residues were arrayed in two adjacent rows or columns of the table, this ratio would be 3.5 : 1. The noise-immunity of the code is therefore high but not optimal.

It should also be noted that Asp, Pro, Glu, and His (residues that play a special role in protein structure, since they interrupt the α-helical segments) are located in adjacent squares in the key in Fig. 10.

Figure 9 indicates the complementary codons. If the nucleotides in the two complementary DNA chains are arrayed in an antiparallel manner ([3], p. 253), as is actually the case, the complementary mRNA codons take the following typical forms:

$$3' \text{ GAC } 5' \text{ (Asp 10),}$$
$$5' \text{ CUG } 3' \text{ (Val 10).}$$

It can be seen from Fig. 9 that, in 24 of 32 cases, the transition to the antiparallel complementary codon is a p → n transition. This is also true of the "parallel" codons, such as:

$$3' \text{ GAC } 5' \text{ (Asp XXXI),}$$
$$3' \text{ CUG } 5' \text{ (Leu XXXI).}$$

If complementary proteins were synthesized in the cell, their structures would have to differ materially. Specifically, if one protein contained an excess of hydrophobic residues, the second would contain an excess of hydrophilic residues. In the limiting case, the first protein would be globular and the second would be fibrillar (see p. 24). However, this is not the case: complementary proteins are not synthesized in the cell, since the mRNA on which the polypeptide chain is assembled is synthesized on only one of the two chains of the dihelical DNA ([3], p. 322).

The correlation that has been found [41, 45, 49] between an amino acid residue and its codon apparently reflects the intimate molecular processes associated with protein biosynthesis.

The amino acid residue, in the form of an aminoacyl adenylate, is attached to a tRNA molecule, which carries an anticodon, i.e., a triplet complementary to the mRNA codon ([3], p. 328). The correlation between the codon and the residue implies a correla-

Table 9. System of Codons

First octet					Second octet					
z = A, G, U, C					z = U, C			z = A, G		
x	y	Amino acid residue	Polarity	n	Amino acid residue	x	y	Amino acid residue	Polarity	n
G	G	Gly	n	6	Cys	U	G	$\begin{cases} -(z = A) \\ \text{Trp } (z = G) \end{cases}$	n	5
G	U	Val	n	5	Ile	A	U	$\begin{cases} \text{Ile } (z = A) \\ \text{Met } (z = G) \end{cases}$	n	4
C	U	Leu	n	5	Phe	U	U	Leu	n	4
G	C	Ala	n	6	Tyr	U	A	–	p	4
C	C	Pro	n	6	Asp-N	A	A	Lys	p	4
A	C	Thr	p	5	His	C	A	Glu-N	p	5
U	C	Ser	p	5	Asp	G	A	Glu	p	5
C	G	Arg	p	6	Ser	A	G	Arg	p	5

tion between the tRNA anticodon and the aminoacyl adenylate. The interaction between the latter and mRNA involves participation of a specific enzyme. It can therefore be surmised that this correlation is dictated by the structure of the enzyme, i.e., that the enzyme has active centers with which the two ligands (the aminoacyl adenylate and tRNA anticodon) interact. We cannot exclude the possibility that the enzyme has allosteric properties (see p. 120), but we as yet know nothing about this problem.

In concluding this chapter, we will consider one other characteristic of the genetic code. Rumer noted that 16 doublets can be grouped into two octets in such fashion that the first octet contains xy doublets that unambiguously define the amino acid residue coded regardless of the ending z, while the second octet contains xy doublets that encode certain residues with purine endings and other residues with pyrimidine endings [51].

Table 9 presents a system of codons that takes into account both the separation into octets and the different coding of polar and nonpolar residues [52].

   Leu, Ser, and Arg are coded by codons of both octets. The
codon sequence in the octets is determined by the sequence of
roots y in the order GUCAG. In the first octet, the prefixes x are
arranged in a sequence that provides separation of the p- and n-
residues (the division is shown by a horizontal line): the order in
this case is GCAU. The codons of the second octet are derived
from those of the first octet by the following rules:

$$G \leftrightarrow G,$$
$$U \leftrightarrow U,$$
$$C \leftrightarrow A$$

for the roots y, and

$$G \leftrightarrow U,$$
$$C \leftrightarrow A$$

for the prefixes x. The polar and nonpolar amino acids are again
clearly differentiated with this substitution pattern.

   The xy doublets of the first and second octets differ greatly
in composition. The first octet contains A only once, while the
second contains C only once. As a result, $(G + C)/(A + U) = 3$
for both the roots and prefixes in the first octet, while
$(G + C)/(A + U) = \frac{1}{3}$ in the second octet.

   The ratio $(C + A)/(G + A) = 1$ for the prefixes in both the
first and second octets, but it equals 3 for the roots in the first
octet and $\frac{1}{3}$ for those in the second octet.

   These differences in the composition of the codons force us
to consider the nature of their interaction with the complementary
anticodons. Here we introduce the concept of the degree of codon–
anticodon complementarity, which is characterized by the number
of hydrogen bonds n between the nucleotides of the prefix x and
root y and the corresponding nucleotides x' and y'. Table 9 gives
the appropriate figures. Proceeding from the fact that G is connected to C by three hydrogen bonds and U to A by two hydrogen
bonds, we find that the values of n in the first octet are 6 and 5,
while those in the second octet are 5 and 4 (averaging 5.5 and 4.5).
It can be assumed that, when n = 6 for the prefix and root xy, the
ending interaction z–z' between the codon and anticodon is of no
great importance, since the xy–x'y' bonding is rather strong and
provides the necessary complementarity. It is for precisely this

reason that the codons of the first octet are independent of their endings. We can conceive of 16 codon—anticodon combinations with which the same amino acid will be incorporated into the protein chain and correspond to any z and z' with constant xy, and hence x'y'. If n = 5 in the first octet, the z—z' interaction can play some role in providing complementarity and the number of codon—anticodon combinations that must be considered is less than 16 but more than 4. It naturally does not follow from the foregoing that all conceivable combinations are found in nature. Nevertheless, it would be interesting to make an experimental determination of the number of different anticodons, i.e., different tRNA's, corresponding to a given amino acid and, of course, of the number of corresponding codons, which is immeasurably more difficult. It is possible that such an experiment would actually demonstrate the existence of a larger number of anticodons and codons for those amino acids of the first octet for which n = 6 and a smaller number for those amino acids for which n = 5.

Mutational substitutions of the endings z in mRNA codons can accordingly differ markedly in their effectiveness in cases where n = 6 and n = 5.

The type of ending z (whether it is a purine or pyrimidine nucleotide) is important in the second octet. At n = 5, the maximum number of codon—anticodon combinations corresponding to the same amino acid in the second octet is 8 (provided that z' is not constant). At n = 4, the z—z' bond must be assumed to be predetermined and the number of combinations is 2.

We have seen that the genetic code is not at all random in character and the characteristics described above indicate that it follows natural rules. A molecular interpretation will be possible after more detailed experimental study of protein biosynthesis. Crick recently considered the properties of the anticodon ending in his theory of "wobbles" [318].

Chapter 5

MACROMOLECULAR PROPERTIES
OF ENZYMES

The enzymatic activity of proteins is governed by the three-dimensional structure of their molecules as integral systems and by the specific chemical and physicochemical properties of the amino acid residues, which interact with the compounds participating in the reaction catalyzed. Braunshtein quite clearly indicated the factors that operate in enzymatic catalysis [53, 53a]:

1. A high degree of affinity between enzyme and substrate, i.e., a high probability of formation of an enzyme—substrate complex, which is equivalent to a sharp rise in reagent concentration (the approximation effect). In other words, the reagent molecules are forcibly drawn to the surface of the larger enzyme molecule.

2. A strict relative orientation of the reagents, coenzymes, and enzyme active center (the orientation effect). In ordinary homogeneous reactions, the probability of strict orientation of three or more interacting molecules is negligibly small.

3. Interaction of the nucleophilic and electrophilic groups of the enzyme active center in the contact area of the complex (synchronous intramolecular acid-base catalysis effects).

4. Activation of the substrate by redistribution of its electron density under the influence of the electrically active groups (the polarization effect).

The process begins with formation of an enzyme—substrate complex, i.e., adsorption of the substrate to the enzyme surface. Enzyme—substrate interactions differ in character: there are electrostatic dipole—dipole and ion—dipole interactions, hydrogen bonds, van der Waals interactions, and hydrophobic and chemical interactions. The net result of these interactions is a decrease in the activation barrier of the reaction and an increase in its absolute rate. The main factor responsible for this is replacement of low-probability, high-order reactions requiring collision of three or more molecules by highly efficient first-order reactions, i.e., the reactions of multifunctional catalysis. The different functional groups exert a "concerted" action on the substrate.

Two fundamental questions can be posed in connection with the foregoing. First, which general properties of an enzyme macromolecule as a physical system provide for realization of the effects enumerated above? Can these effects be interpreted by considering the macromolecule to be a static rigid system, whose specificity is governed solely by the set of amino acid residues at its surface?

Secondly, are the effects discussed above sufficient to account for enzymatic catalysis? Do they quantitatively lead to the observed decrease in activation energy, or are there also other physical mechanisms, particularly mechanisms that furnish additional energy to the reagents? There are as yet no complete answers to either question. It has not proved possible to make a strict theoretical calculation of enzymatic activity in any specific case, despite the fact that the chemical nature of enzyme action has been determined in a number of reactions. The search for answers to the above questions consequently leads us into the area of hypothesis and scientific speculation. As has already been noted, protein molecules are not completely ordered (see p. 13). A larger or smaller number of the amino acid units are in a state corresponding to a disordered secondary structure. Ordering and disordering processes, particularly α-helix formation and breakdown, can therefore occur in such a molecule. Everything that we know about the properties of macromolecules forces us to assume an enzyme molecule to be a fluctuating, dynamic system. This raises doubts as to the reliability of the results obtained in investigations of protein structure. X-ray diffraction studies have been made with crystalline proteins. It is possible that the crystal structure of a protein differs from its structure in solution, but there are data indicating that no such differences exist [54]. In particular, the degrees of α-helicity determined in crystals and in solution agree [55]. We therefore need not raise this sort of objection, although the problem has not been conclusively solved.

Determination of the conformational properties of enzymes in solutions, e.g., the degree of α-helicity by spectropolarimetry (see Chapter 8), obviously provides information on some average state of the fluctuating macromolecule. The difference in the free energies of the ordered and disordered states is small: ΔF is of the order of 1 kcal per mole of residues. At the same time, ΔH and $T\Delta S$ are far larger, but compensate for one another. It is for precisely this reason that the conformational fluctuations may be substantial [56]. A number of arguments can be advanced to support the hypothesis that enzyme macromolecules actually fluctuate. Linderstrøm-Lang studied the exchange of the hydrogen of peptide groups (CO—NH) for deuterium [56]. If there were only an ordered structure with intramolecular hydrogen bonds, deuterium exchange would be greatly impeded. However, this process occurs at temperatures substantially below the denaturation point of the protein, at which the hydrogen bonds are ruptured. Study of enzyme—substrate complexes provides indirect information on the existence of fluctuations, e.g., the dynamic nature of enzymes. Such a complex usually has a more rigid structure than the free enzyme, as is demonstrated by the fact that complexes exhibit a lower trypsin-digestibility than free enzymes [57-59]. It can be surmised that the binding of the substrate by the enzyme reduces its fluctuational lability.

Direct physical techniques for studying conformational fluctuations should be kinetic rather than structural. It might be most promising to undertake research on the polarization of the fluorescence of both aromatic amino acid residues present in a protein and fluorescent molecules attached to it (see [8]).

The role of structural fluctuations can be twofold. On the one hand, the existence of such fluctuations can provide for development of the proper enzyme—substrate conformation. A hypothesis of this type will be discussed below. On the other hand, it is conceivable that fluctuations play a material energetic role as the factor responsible for accumulation of additional energy in the chemical bonds of the substrate attached to the enzyme. This alternative is far more problematical. Nevertheless, it requires analysis and discussion.

Kirkwood and Shumaker considered the fluctuations in the electrical charges on the ionogenic groups of a protein rather than its conformational fluctuations. They concluded that the enzyme—substrate interaction, which is governed by these fluctuations, can acquire additional electrostatic energy [60]. They discussed only ion—dipole interactions. However, this article, like that discussed later (see p. 96), did not explain the general nature of enzymatic activity. The approach was oversimplified and the very existence of charge fluctuations cannot be regarded as proved.

An enzyme globule is a rather dense, compact structure differing materially from an open, randomly folded structure. In this sense, a globule resembles a drop of liquid, in which the molecules are densely packed. Ptitsyn and Eizner developed a convincing theory to account for the globule—random coil transition, proceeding by analogy with the cooperative process of liquid evaporation [61]. Protein denaturation may have the character of such a transition.

The fluctuations in density in a drop of liquid are a superpositioning of sound waves. Einstein represented the density fluctuations in a liquid as a Fourier series [62]. Mandel'shtam later showed that these Fourier components describe real sound waves and suggested a method for studying them experimentally; the first investigations were conducted by Gross (see [63], Chapters 5 and 6). The set of density fluctuations in a liquid is similar to the acoustic branch of vibrations in a solid, but the upper frequency limit is determined by the inverse relaxation time τ^{-1} of the liquid.

At the same time, it must be kept in mind that the drop model under discussion is very arbitrary. There is equal justification for speaking of the acoustic vibrations of a solid simulating an enzyme, i.e., of phonon propagation.

Do acoustic (hyperacoustic) vibrations play any part in enzymatic activity? One can hypothesize that the vibration energy accumulates at the site of substrate attachment, i.e., in the active center of the enzyme [64]. Enzymatic activity should then depend directly on the globule size. Until recently, this was contradicted by a number of data indicating that an enzyme retains its activity after a substantial portion of the peptide chain had been split off from it [65, 66]. However, contrary results have lately been obtained: it was found that the protein fragments previously studied

contained an admixture of undecomposed enzymes [67, 68]. This
eliminates an important objection to the hypothesis in question,
but it has not yet been confirmed. It has not received proper theoretical treatment and there are no techniques for verifying it experimentally. Study of the kinetic processes within a protein globule by spectroscopy, from the absorption and dispersion of electrical oscillations, and by nuclear magnetic resonance is greatly
hampered by the presence of an aqueous medium (see [319, 320],
among other sources).

Nevertheless, there are some grounds for considering the
vibrations of the drop model to be responsible for the specific
globule interaction in enzymes having a quaternary structure (see
p. 116).

Shnol' reported observing synchronous fluctuations in the
enzymatic activity of actin, myosin, and actomyosin in solution
[69]. He suggested that the protein conformation varies, a process accompanied by a change in the hydrophilic—hydrophobic
properties of the globule surface, which causes rearrangement of
the water structure. The "hydrophilic—hydrophobic waves" prop-
agating in the water cause coupling of the molecular vibrations
and result in their becoming synchronized throughout the entire
solution volume. The vibrations must have acoustic frequencies.
These results, although still partial, may be correlated with the
drop model of the globule (see p. 116). Another conceivable mechanism of vibration — energy accumulation in the enzyme — substrate complex is based on consideration of optical vibrations
with frequencies in the infrared region rather than acoustic vibrations. We will proceed from the theory of thermal unimolecular
decomposition. Let us imagine a chemical bond incorporated into
a complex system of other bonds. The system as a whole undergoes thermal fluctuations. There is a finite probability that energy sufficient for rupture will accumulate at a given bond. This
process is obviously impossible for an isolated bond. It would appear that complexing of the substrate and enzyme can make it possible for energy to accumulate at the substrate bonds.

This notion is very attractive, but, unfortunately, erroneous.
An appropriate theory of unimolecular decomposition was devised
by Slater [70, 71] (see also the article by Obreimov [72]). Calcu-

lations have shown that the accumulation of vibrational energy has
no effect on the activation energy necessary for bond rupture, af-
fecting only the pre-exponential multiplier, which has the sense of
the average vibrational frequency of the system. The expression
for the rate constant of unimolecular decomposition has the
form:

$$k = \bar{v}e^{-E_a/kT}. \tag{12}$$

The average vibration frequency in the enzyme—substrate complex cannot differ materially from the vibration frequency of the
bonds in the substrate molecule.

Finally, we might mention predissociation. The spectrum
of vibrational energy levels is almost continuous in complex molecules, which naturally include proteins [73]. If energy can be
transferred from the vibrational continuum of the protein to the
energy levels of the substrate, predissociation occurs and the
activation barrier is reduced. No research at all has been done
on this possibility.

We have seen that hypotheses regarding the physical processes of energy accumulation in the active center of an enzyme
are very problematical. There are now far better grounds for assuming that an enzyme is not a reservoir for excess thermal energy and that the activation energy is reduced directly and not because the energy of the complex has increased. Study of the
thermodynamics of enzymatic reactions indicates that the entropic term of the free energy plays a large role (see, for example,
[74]). Substantial changes in entropy occur during formation of
the enzyme—substrate complex and decomposition of the enzyme—
product complex.

The following conclusion can be drawn from the foregoing.

1. Conformational fluctuations apparently occur in an enzyme macromolecule. These can be regarded as acoustic vibrations of a drop model and they may play an important role in enzymes with a quaternary structure.

2. There is at present, however, no experimental or theoretical confirmation that fluctuational energy accumulates in the
enzyme—substrate complex.

3. Conversely, conformational fluctuations are apparently very important in the stereochemical (as opposed to energetic) sense.

The following chapter treats the latter aspect of their role.

Chapter 6

INDUCED STRUCTURAL
CORRESPONDENCE OF
ENZYME AND SUBSTRATE

Treatment of the enzyme macromolecule as a fluctuating dynamic structure permits us to postulate a special interaction between enzymes and substrates or other molecules. It is possible that certain protein conformations bind a substrate more effectively than others. This is also true of the conformations of the substrate, which usually has greater conformational lability than a globular protein macromolecule. If the free energy of such an interaction is sufficiently high, it should raise the probability that definite enzyme and substrate conformations will be realized and maintained in the enzyme—substrate complex. The substrate can consequently select the enzyme conformation and vice versa. In 1950, Karush attributed the almost universal capacity of serum albumin to bind different compounds to its "configurational adaptability" [75].

It is obvious that, if such selection of conformations actually occurs, it results from the real forces of the enzyme—substrate interaction. These forces increase with the number of atomic groups in the substrate that are in contact with functional groups of the protein. The interactions themselves are very diverse, because of the wide range of amino acid residues: they include chemical donor—acceptor and ionic bonds, electrostatic ion—dipole and dipole—dipole forces, hydrogen bonds, and hydrophobic interactions of nonpolar groups in an aqueous medium. Contact between interacting groups entails organization of structural correspondence between the surfaces of the substrate and protein molecules.

It can now be considered solidly established that molecular structural correspondence plays the most important role in molecular
biology. Structural correspondence of the complementary chains
occurs in the DNA double helix. The same three-dimensional
complementarity obtains between one of the DNA chains and the
mRNA synthesized on it (as on a template), between the mRNA
codon in a polysome and the tRNA anticodon that carries the amino
acid, and between the DNA and histones or protamines in the
chromosomes. Structural correspondence of the key-and-lock
type is apparently especially manifest in antigen—antibody interactions in immunochemical processes. There are grounds for de-
ducing that structural correspondences govern the formation of
supermolecular structures during morphogenesis [76], such phenomena as chromosomal synapsis, etc. In short, structural correspondence can scarcely but be the most important and universal
type of biomolecular mechanism ([3], Chapter 8). The structural
correspondence between an enzyme and its substrate can be defined as dynamic and induced, in contrast to, for example, the
static correspondence between an antigen and antibody. As has
now been proved, an antibody synthesized in the cell acquires a
predetermined structure complementary to that of its antigen (see,
for example, [77, 78]). It must be noted that the idea of induced
structural correspondence was first advanced in immunochemistry.
Pauling believed that an antigen molds the structure of the flexible
antibody molecule, organizing such correspondence [79]. This
theory was not confirmed immunochemically, but good ideas never
fail to bear fruit. Induced structural correspondence is apparently an important factor in enzymology. Organization of structural
correspondence by selection of protein conformations should be
manifested in the specific thermodynamics of enzymatic reactions.
The entropy of the protein can vary during this process. There
should be a decrease in the entropy of the residues participating
in formation of the active center, but it can be partially or completely compensated for by an increase in the entropy of other
areas of the globule. Vaslov and Doherty reported conformational
effects during the binding of substrates and competitive inhibitors
by chymotrypsin [80]. The literature contains similar observa-
tions. The change in entropy during formation of the enzyme—
substrate complex apparently indicates an actual change in protein
conformation. This has given rise to the concept of the dynamic
"rack": the substrate on the enzyme surface is in a stressed, ex-

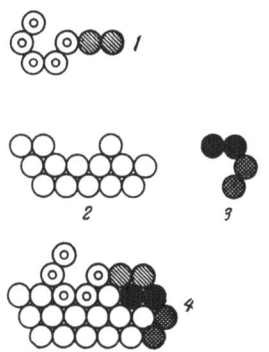

Fig. 11. Diagram of formation of enzyme—substrate complex (after Balandin). 1) Substrate; 2) apoenzyme; 3) coenzyme; 4) complex.

tended, or folded state, so that the activation energy of the process being catalyzed is reduced [74]. A pronounced effect of this type can occur only when the elastic forces in the deformed polypeptide chain are large. The change in free energy per residue is small, being of the order of kT, i.e., 600 cal/mole, and cannot markedly reduce the activation barrier. However, a set of interacting residues participates in the process and the free energy of their cooperative deformation may accumulate in the substrate molecule. We must again return to the previously considered fluctuation theories. No investigations at all have been conducted in this area; the experimentally observed integral thermodynamic effects provide only indirect information.

The structural correspondence of enzymes and substrates was considered in detail by Balandin, who developed the so-called multiplet theory of catalysis [81-83]. The notion of a structural correspondence between the catalyst and the reagents is the starting point for this general theory of heterogeneous catalysis. Figure 11 shows Balandin's diagram of an enzyme—substrate complex. Formation of a compact structure from the apoenzyme, coenzyme, and substrate leads to a corresponding reduction in the activation barrier. These are naturally only qualitative considerations. They take no account whatsoever of the cooperative dynamics of the process, being limited solely by geometric factors.

As has already been pointed out, the conformation of a substrate molecule should be altered when it is brought into structural correspondence, provided that it is flexible, i.e., capable of rotary isomerization. Phenomena of this type have not been studied at all, although they are exceptionally interesting. It is obviously difficult to detect entropic effects in the substrate against the background of the far greater effects in the enzyme, which contains an immeasurably larger number of bonds. Direct structural investigative techniques are required.

Koshland further developed the hypothesis of induced structural correspondence in enzymatic catalysis.

This hypothesis was devised principally to account for the specificity of enzymes that catalyze bond—transfer reactions of the type

$$(B-X) + Y + E \rightleftarrows (B-Y) + X + E,$$

where E is an enzyme. The old notion of a static key-and-lock system attributed the specificity of enzymes to the rigidity of their structure, which results in attraction of a definite substrate molecule and steric repulsion of somewhat different analogs. Thus, substitution of a methoxyl group for a hydroxyl or of CH_3 for H can convert a substrate to a competitive inhibitor. Koshland showed that a number of phenomena contradict so simple a model [84]. Let us consider these phenomena.

Water and small hydroxyl-containing molecules are not reactive in hydroxyl—transfer reactions catalyzed by phosphorylases and kinases. On the other hand, larger hydroxyl-containing molecules serve as substrates. Some compounds adsorbed by the active center of an enzyme are nonreactive or poorly reactive. The catalytic groups of the enzyme have little or no effect on such compounds, despite the fact that similar adsorbed substances are reactive.

Some small molecules are not attached to the enzyme surface, while their larger analogs are bound. Thus, 2-deoxyglucose is a hexokinase substrate, while 2'-methoxyglucose is neither a substrate nor an inhibitor. It might be assumed that replacement of the OH by OCH_3 makes attachment impossible; however, both N-methylglucosamine and N-acetylglucosamine, which contain large groups, are competitive inhibitors.

There are many other examples of lack of reactivity in small molecules with a structure similar to that of a bulkier normal substrate. Phosphotransacetylase acts on acetate, propionate, and butyrate, but not on formate, while β-glucosidase acts on glucosides but not on 2-deoxyglucosides. Koshland formulated the hypothesis of induced structural correspondence, which gives a qualitative interpretation of the phenomena described above and others similar to them, in the form of three postulates [84, 85]:

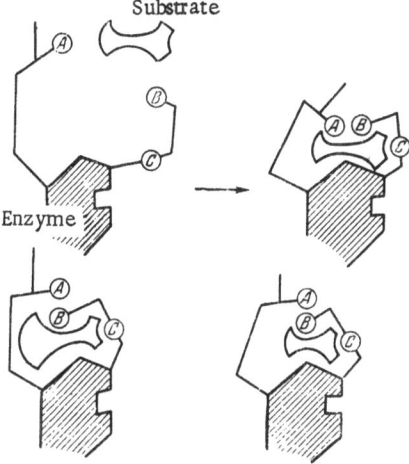

Fig. 12. Diagram of interaction of an enzyme with a substrate and its analogs (after Koshland).

1. The substrate causes material changes in enzyme geometry, since it penetrates into the active center.

2. Precise orientation of the catalytic groups is necessary for catalytic action.

3. The substrate induces the proper orientation by the changes it produces in enzyme geometry.

Figure 12 is a diagram of this process. It actually represents a qualitative interpretation of the anomalies enumerated above. Since the enzyme is a flexible system, small molecules may fail to be sorbed or be nonreactive, while larger molecules are functional. An important feature of Koshland's model is differentiation of the ability of an enzyme to bind a substrate molecule and its ability to catalyze substrate transformations. The specificity and catalytic efficiency of an enzyme are interrelated in this sense, but they have different mechanisms. A reaction occurs only with the correct relative positioning of the sorptive and catalytic centers with respect to the substrate molecule. Koshland illustrates this with β-amylase [86]. This enzyme acts on the terminal groups of amylose but not on the other glucoside

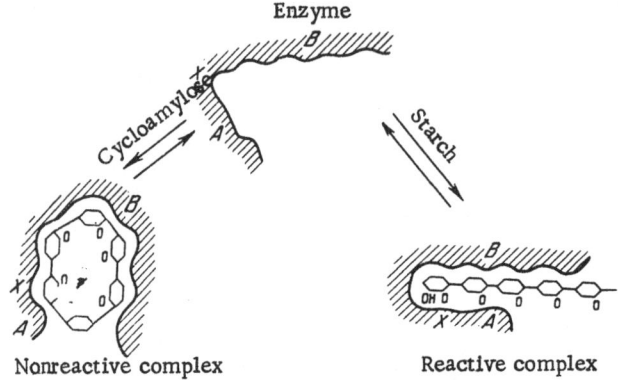

Fig. 13. Diagram of action of β-amylase (after Koshland).

bonds of the polysaccharide. Cycloamyloses are competitive inhibitors of the enzyme. This is explained by the diagram in Fig. 13.
The reaction occurs only when the effective groups of the enzyme
(sorptive and catalytic, designated by A, B, and X) have a definite
three-dimensional arrangement.

The second problem is to substantiate the second postulate
of the hypothesis, even though only semiquantitatively. Can approximation and precise relative orientation of the reagents cause
a large increase in reaction rate? Koshland made appropriate
rough calculations [87].

Let us assume that we are considering a reaction in an aqueous solution:

$$A + B \rightleftarrows C + D.$$

In order for the reaction to occur, it is necessary that A and B
molecules collide. If we employ the lattice model of a liquid (see,
for example, [88]) and assign the coordination number γ (Koshland
uses $\gamma = 12$), the molar concentration of A—B pairs is represented by the quantity

$$\frac{[A] \cdot [B] \cdot \gamma}{55.5},$$

where [A] and [B] are the molar concentrations of the reagents
and 55.5 = 1000 : 18, and is the molar concentration of water.

The reaction rate in the absence of an enzyme is

$$v_0 = k_0 \, [A] \, [B]. \tag{13}$$

In the presence of an enzyme, the rate is altered and is determined by the enzyme concentration, i.e.,

$$v_e = k_E \, [E]. \tag{14}$$

It can be roughly assumed that

$$k_0 \sim \frac{\gamma}{55.5}, \quad k_E \sim [E].$$

Hence,

$$\frac{v_e}{v_0} \approx \frac{[E] \, 55.5}{[A] \, [B] \, \gamma}. \tag{15}$$

If the concentration of enzyme E is small, $v_e \ll v_0$. However, we have taken into account neither contact nor orientation. Assuming that the functional groups R, S, and T of the enzyme are located inside the lattice cells, we obtain an expression for the concentration of enzyme—substrate complexes of the type shown in Fig. 14a:

$$\frac{[A] \, [B] \, \gamma \, [R] \, (\gamma - 1) \, [S] \, (\gamma - 2) \, [T] \, (\gamma - 1)}{(55.5)^4}, \tag{16}$$

whence

$$\frac{v_e}{v_0} \approx \frac{[E] \, (55.5)^4}{[A] \, [B] \, [R] \, [S] \, [T] \, \gamma \, (\gamma - 1)^2 \, (\gamma - 2)}. \tag{17}$$

Koshland introduces the additional factors f, which represent the fraction of the total number of molecules in nearest-neighbor positions (in which they can react), and θ', which represent the relative orientation (Fig. 14b). Then,

$$\frac{v_e}{v_0} \approx \frac{[E] \, (55.5)^4 \, \theta'_B \theta'_R \theta'_S \theta'_T}{[A] \, [B] \, [R] \, [S] \, [T] \, \gamma \, (\gamma - 1)^2 \, (\gamma - 2) \, f_{AB} f_{BR} f_{BS} f_{AT}}. \tag{18}$$

Assuming that the values of f are of the order of $1/\gamma$, and those of θ' are of the order of 10, Koshland showed that the mutual contact and orientation effects can increase the reaction rate by as much

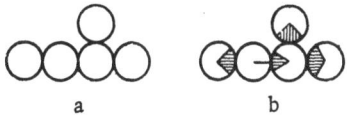

a   b

Fig. 14. Diagram of contact between reacting groups (a) and of contact with mutual orientation (b).

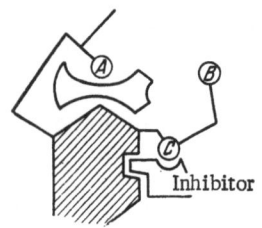

Fig. 15. Diagram of action of inhibitor.

as 18 orders of magnitude. This calculation is naturally not a quantitative proof, but it is of illustrative value.

In a later work, Koshland proposed a graphic model for the action of competitive and noncompetitive inhibitors and of activators or hormones [89]. This model is shown in Fig. 15. An inhibitor is competitive if group B is essential for attachment of the substrate and noncompetitive if group B is catalytic. Koshland considers a hormone to be a molecule that induces proper packing of the polypeptide chain even in the presence of an inhibitor.

Finally, in one of his most recent articles, Koshland gave more detailed consideration to the behavior of the substrate in an enzyme—substrate complex [90, 91]. The hypothesis of induced structural correspondence was thus formulated qualitatively and illustrated by graphic diagrams. However, it has not received any quantitative theoretical substantiation and the experimental confirmation cannot be considered exhaustive.

First of all, it is necessary to prove the occurrence of conformational changes in the protein during its interaction with its substrate, inhibitors, and other effectors. Secondly, if such changes are detected, it must be demonstrated that they actually cause the observed decrease in activation energy.

There are a number of indirect proofs of conformational changes in proteins. We have already mentioned the thermodynamics of enzymatic reactions. Clearer confirmations of such changes are enumerated in an article by Koshland, Yankeelov, and Thoma [86]. Some enzymes become more rigid in the presence of substrates (see p. 42), while others become more labile and are more readily denatured on heating [92]. Substrates induce dissociation of glutamate dehydrogenase [93] and hexokinase [94]. The substrate alters the reactivity of the enzyme, so that, for example, a substrate intensifies the iodination of penicillinase [95]. A qualitative interpretation of such phenomena, essentially based on the Koshland model, is given by Ling [96]. Numerous additional data of this sort are presented in the survey by Jencks [97]. In some cases, on the other hand, the substrate stabilizes the enzyme structure, making it more resistant to denaturation. Thus, removal of adenosine triphosphate (ATP), which can be regarded as a substrate of G-actin, causes denaturation of the latter, as is shown by changes in the optical rotation of this protein and an increase in its capacity for proteolytic decomposition [98]. The list of similar proofs of conformational changes in enzymes could be made much longer, but there is no need to do so (see also the surveys by Koshland [99] and Braunshtein et al. [53a]).

Optical and spectroscopic methods provide more direct information. Optical activity will be considered below. Changes in the spectrum of chymotrypsin during its interaction with a substrate were observed by Wooton and Hess [100], who thought that they were caused by conformational changes in the chymotrypsin molecule. There are a number of other reports containing similar information (see also [101]).

Conformational transitions in proteins have been observed in studying fluorescence intensity and polarization. Either the aromatic amino acid residues of proteins or luminophores (dye molecules) adsorbed to proteins fluoresce (see the exhaustive bibliography in the article by Fasman et al. [102]).

Yankeelov and Koshland recently conducted a detailed investigation of the conformational changes induced in phosphoglucomutase (PGM) by the substrate glucose-6-phosphate and its analogs [103]. They employed a number of indirect techniques which, in aggregate, provided sufficiently convincing proof that such changes occur. This work was quite instructive and we will therefore discuss it at greater length.

First of all, it was established that the substrate increases the reactivity of the protein SH groups by 20% with respect to alkylation by iodoacetamide. There is a corresponding rise in the rate at which PGM is inactivated by this compound. Other substances, particularly glycerophosphate (a substrate analog), have no such effect. The authors attribute this result to exposure of some portion of the SH groups (PGM contains a total of six) as a result of a conformational change caused by the substrate. These groups are inaccessible to the reagent in the free enzyme but are exposed in the enzyme—substrate complex. In this connection, it

is validly noted that a change in the reactivity of the amino acid
residues of a protein can also occur without induction of structural
correspondence, simply as a result of steric hindrance to the reaction produced by the presence of the substrate. There should be a
decrease in reactivity in this case, while an observed increase corresponds to a change in the conformation of the protein molecule
as a whole. Similar phenomena were previously established for
other enzymes. The methionine residue in ribonuclease is nonreactive at pH 6 but becomes reactive when the macromolecule is un-
folded by raising the pH to 8 or exposing it to urea [104]. The same
residue loses its reactivity in chymotrypsin under the action of
various agents: photooxidation, hydrogen peroxide, iodine chloride,
iodoacetamide, or iodoacetic acid [105]. It must be noted, however,
that the data obtained in these investigations are not very conclusive, since they do not deal with factors specific for the active center.

  Yankeelov and Koshland detected changes in the optical properties of PGM during its interaction with a substrate. Thus, they
observed a change in the differential absorption spectra in the 260
to 300 mμ region and in the fluorescence spectra (a maximum at
330 mμ). These indicate changes in the neighborhood of the tyrosine and tryptophan residues, which are aromatic residues with
characteristic long-wavelength spectral bands. The information
obtained is indirect, but the results become conclusive when con-
joined with chemical data.

  The active center of PGM contains phosphoserine. Dephosphorylation of PGM markedly alters all the aforementioned properties. The events described above are illustrated schematically in
Fig. 16. The conformation of the enzyme molecule changes during
the interaction with the substrate; one SH group is exposed, the X,
Y, and Z groups are masked, and the W group moves from a hydrophilic to a hydrophobic environment. The W, X, Y, and Z groups
may correspond to Trp, Lys, Met, and Cys residues. The diagram
shows that the presence of the substrate masks the Y and Z groups
as effectively as the X group. If this interpretation is correct, it
gives rise to an important consequence. A decrease in the reactivity of certain groups does not necessarily mean that the protected
residue is in the active center, in direct contact with the substrate.
If the substrate causes a change in the reactivity of many different

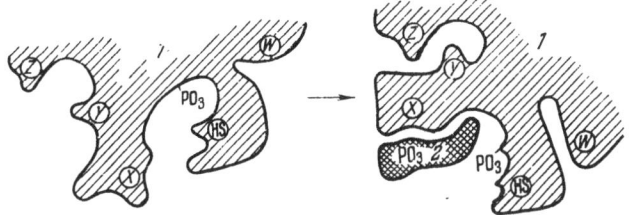

Fig. 16. Diagram of conformational transformation in phosphoglucomutase (after Yankeelov and Koshland). 1) Enzyme; 2) substrate;
X, Y, Z, W) Functional groups of enzyme.

groups, this can be regarded as solid confirmation of the hypothesis that conformational transformations take part in enzymatic
catalysis.

Study of allosteric enzymes provides important information
on the changes in the higher structural levels of a protein during
its interactions with substrates, inhibitors, and activators (see
Chapters 12 and 13).

There are thus a number of data indicating that conformational changes, i.e., changes in the higher structural levels of the
protein, occur when an enzyme interacts with a substrate or other
effector. However, the data presented in this chapter are, for the
most part, indirect and incomplete proofs of the theory of induced
structural correspondence. There is no direct proof that the observed conformational changes involve both the active center of
the enzyme and the structure of the macromolecule as a whole,
nor that these changes lead to structural correspondence between
the enzyme and the substrate.

The hypothesis of induced correspondence is obviously of
great significance for interpretation of enzymatic catalysis. Further research on the conformational properties of enzymes is
therefore a very pressing task. Direct investigations by physical
techniques must be undertaken and a quantitative theory must be
devised, something that no one has yet attempted.

The most direct techniques for studying the conformational
properties of proteins are those of spectropolarimetry. Investigations of this type will be discussed below.

In conclusion, it should be pointed out that one enzyme—inhibitor complex has been investigated by x-ray diffraction analysis. Crystalline complexes of lysozyme with certain inhibitors have been studied [106] (see also [12]). The six-angstrom resolution was insufficient to show any detail, but it proved possible to establish that there is a spatial correspondence between the enzyme and inhibitor molecules.

At the Seventh International Congress on Crystallography in Moscow in July, 1966, Phillips reported on the results of further x-ray studies of lysozyme with greater resolution [107]. Consideration of hydrophobic interactions made it possible to explain the development of a definite tertiary structure in this protein, proceeding from the known primary structure synthesized unit by unit on mRNA. Phillips made a detailed study of the nature of complexes of the enzyme and substrate analogs. He demonstrated that lysozyme catalyzes the hydrolytic decomposition of carbohydrates, particularly chitin. Oligomers of the chitin type, trimers and hexamers form complexes with lysozyme by entering a deep fissure in the globule. Phillips established the specific interactions between the functional groups of the carbohydrate and the amino acid residues, and showed which bonds in the carbohydrate are ruptured and how they are broken.

The interaction of lysozyme with oligomeric carbohydrates is effected by the structural-correspondence principle, but it is not accompanied by any material conformational changes. Thus, according to Phillips's estimate, the displacement of the Trp residue (No. 62) does not exceed 0.75 Å. Moreover, modeling has shown that the conformational rearrangement of the substrate must be considerable. The fourth unit in the hexamer cannot enter the fissure, being in the "chair" conformation normal for carbohydrates, and a "semichair" conformation should be obtained. This first direct investigation of an enzyme—substrate complex thus confirmed Koshland's hypothesis only in general form. In principle, induced structural correspondence of an enzyme and substrate can be achieved by preferential conformational rearrangement of the substrate in some cases and of the enzyme in others.

Chapter 7

OPTICAL ACTIVITY

The optical activity of molecules is extremely sensitive to changes in their structure, particularly to conformational transformations, and depends to a large extent on intermolecular interactions. The reason for this high sensitivity is quite clear, lying in the difference in the phases of a light wave at different points in an asymmetric molecule. In studying the optical activity of a molecule, we are thus investigating the interference of light waves in it [63].

Rotation of the plane of polarization of light by a compound means that it has a circular birefringence, i.e., that waves circularly polarized to the right and left propagate through it at different speeds (Fresnel). Actually, a linearly polarized wave can split into two circularly polarized waves, left- and right-handed. If one of these is propagated faster than the other, the plane of polarization is rotated by an angle [63]

$$\varphi = \frac{\pi}{\lambda}\,(n_l - n_r)\,l \quad \text{rad,} \tag{19}$$

where λ is the wavelength of the light in a vacuum, n_l and n_r are the refractive indices for the left-handed and right-handed waves, respectively, and l is the path length of the light in the optically active medium. The specific rotation of a compound in solution is designated by the quantity

$$[\alpha] = \frac{180}{\pi} \cdot \frac{10}{C} \cdot \frac{\varphi}{l} \quad \text{deg} \cdot \text{dm}^{-1} \cdot \text{g}^{-1} \cdot \text{cm}^3, \tag{20}$$

where C is the concentration of the compound in $\text{g} \cdot \text{cm}^{-3}$. The molecular rotation equals

$$[M] = \frac{M}{100} [\alpha], \qquad (21)$$

where M is the molecular weight.

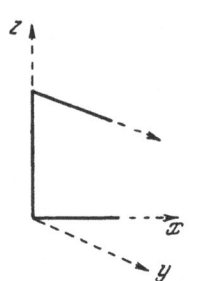

Fig. 17. Model of an optically active molecule (after Kuhn).

The classical theory of this phenomenon, which is based on the model of the electron as a harmonic oscillator, shows that three conditions must be satisfied for optical activity to appear. First, the distance between different electrons in the molecule should not be negligibly small in comparison with the wavelength of the light. In other words, the phase difference of a light wave at different points in the molecule, which we have already mentioned, must be taken into account. Secondly, the electrons should interact with one another. Finally, the molecule as a whole should have neither a plane nor a center of symmetry. The simplest classical model of an optically active molecule, proposed by Kuhn, is accordingly two interacting electrons moving some distance apart in different directions, say along the x and y axes in Fig. 17 [63, 108].

The rigorous quantum-mechanical theory of molecular optical activity is based on the theory of perturbations. It deals with the system formed by a molecule and the electromagnetic field of a light wave, taking into account the phase difference of the wave at different points in the molecule. For a solution with the refractive index n, Equation (19) takes the form

$$\varphi = \frac{4\pi^2 \nu^2}{c^2} 4\pi N_1 \frac{n^2 + 2}{3} \beta l, \qquad (22)$$

where N_1 is the number of active molecules per unit volume. Equations (20)-(22) consequently yield

$$[M] = \frac{288\, \pi^2}{c^2} \nu^2 N_A \frac{n^2 + 2}{3} \beta, \qquad (23)$$

where N_A is the Avogadro number. The molecular parameter β has the form

$$\beta = \frac{c}{3\pi h} \sum_j \frac{\mathrm{Im}\,(p_{0j}\mu_{j0})}{\nu_j^2 - \nu^2}. \qquad (24)$$

Here the symbol Im means that we use the imaginary component of the product of the matrix elements of the electric dipole moment p_{0j} and the magnetic dipole moment μ_{j0}, which correspond to transitions between the ground level 0 and the excited level j; ν_j is the frequency of these transitions, and ν is the frequency of the incident light. Rosenfeld derived Equation (24) [63, 109, 110].

It follows to compare the expression for molecular refraction

$$\frac{n^2-1}{n^2+2}\frac{M}{\rho}=\frac{4\pi}{3}N_A\alpha, \tag{25}$$

with Equation (23); here, ρ is the density of the compound and α is the molecular polarizability. The quantum-mechanical expression for polarizability, which is comparable to the expression for β, has the form

$$\alpha=\frac{2}{3h}\sum_j\frac{\nu_j p_{0j}^2}{\nu_j^2-\nu^2}. \tag{26}$$

The classical analog of Equation (26) is

$$\alpha=\frac{e^2}{4\pi^2 m}\sum_j\frac{f_j}{\nu_j^2-\nu^2}, \tag{26a}$$

where e and m are the charge and mass of the electron, and f_j is the strength of the oscillator. It follows from comparison of Equations (26) and (26a) that

$$f_j=\frac{8\pi^2 m\nu_j}{3e^2 h}p_{0j}^2. \tag{27}$$

The dispersion formulas (24) and (26) are valid for the region of frequencies ν remote from frequencies ν_j, which correspond to the natural absorption bands. There is also a material difference between Equations (24) and (26). The sum of the numerators in Equation (26) is positive, while that in Equation (24), which is the sum of the so-called rotatory strength

$$R_{0j}=\text{Im}\,(p_{0j}\mu_{j0}), \tag{28}$$

equals zero, i.e.,

$$\sum_j R_{0j} = 0.$$

Actually, according to the rules of quantum mechanics,

$$\sum_j \mathrm{Im}\,(p_{0j}\mu_{j0}) = \mathrm{Im}\sum_j p_{0j}\mu_{j0} = \mathrm{Im}\,(p\mu)_{00} = 0,$$

since $(p\mu)_{00}$ is a real number. It is also readily shown that all values of R_{0j} revert to zero when a plane or center of symmetry is present [110].

Let us evaluate the order of magnitude of the rotatory strength R_{0j}. The magnetic moment of the electron is $0.93 \cdot 10^{-20}$ erg $\cdot$ G^{-1}. The order of magnitude of p_{0j} is 1 D, i.e., 10^{-18} CGSU. As a result, R_{0j} is of the order of 10^{-38} CGSU.

Rosenfeld's formula does not permit direct calculation of the optical activity, since this requires a knowledge of the complete sets of molecular wave functions and energy levels. Approximation methods for calculating β have therefore been devised.

Let us separate the molecule into individual groups, between which there is no electron exchange. For a protein, these might be individual peptide groups and amino acid radicals. We then represent β as the sum

$$\beta = \beta_1 + \beta_2 + \beta_3 + \beta_4,$$

(29)

where β_1 is the contribution made by an individual group, equalling zero if the group is symmetric [111]; β_2 are the single-electron terms produced by "mixing" of the electric and magnetic dipole transitions in a single group in the asymmetric field of the remaining groups; β_3 is produced by "mixing" of the magnetic dipole transition of one group with the electric dipole transition of another group; β_4 results from simultaneous dipole excitation of different groups. The expression for β_4 was obtained quantum-mechanically by Kirkwood [112], but it has been shown that it can also be derived by purely classical means [63, 113]. In deriving β_4, it is necessary to consider the electrostatic dipole—dipole interaction of the anisotropically polarized groups in the molecule. We obtain

$$\beta_4 = \frac{1}{6} \sum_{a, b}' (\alpha_{a1} - \alpha_{a2})(\alpha_{b1} - \alpha_{b2})(r_{ab}[a, b])(aT_{ab}b), \qquad (30)$$

where a and b are the group numbers (assuming that the groups are axially symmetric); a and b are unit vectors directed along the group axes; α_1 is the group polarizability along the axis, α_2 is the group polarizability in the perpendicular direction, and r_{ab} is the distance between the groups. The quantity $(a\,T_{ab}b)$ describes the dipole—dipole interaction:

$$(aT_{ab}b) = \frac{1}{r_{ab}^3} \left\{ (ab) - \frac{3\,(ar_{ab})\,(br_{ab})}{r_{ab}^2} \right\}. \qquad (31)$$

The term β_3 is small, its value being at least an order of magnitude below the observed value of β. On the other hand, β_2 and β_4 are of the same order as β [111].

We can clarify the derivation of Equation (30) with a simple example [114]. Let the molecule be a dimer, i.e., consist of two separate particles a and b. An electrostatic interaction with the energy

$$V_{ab} = (p_a T_{ab} p_b). \qquad (32)$$

occurs between the dipoles with the moments $\mathbf{p}_a$ and $\mathbf{p}_b$ induced in the particles. The effective field acting on particle a is

$$E'_a = E - T_{ab} p_b; \qquad (33)$$

that acting on particle b is

$$E'_b = E - T_{ab} p_a, \qquad (34)$$

where

$$p_a = \alpha_a E'_a, \quad p_b = \alpha_b E'_b, \qquad (35)$$

and α_a and α_b are the polarizabilities of the particles.

Let the particles be rods arranged in the manner shown in Fig. 18, i.e., with their centers lying on the z axis at a distance r from one another, while the particles themselves lie in planes parallel to xy at an angle 2θ. The particles are polarized only

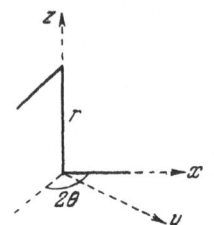

Fig. 18. Model of optically active dimer.

along their axes (parallel to the rods). In this case,

$$(r_{ab}[ab]) = r\sin 2\theta,$$
$$(aT_{ab}b) = \frac{\cos 2\theta}{r^3} = T \tag{36}$$

and, according to Equation (30),

$$\beta = \frac{1}{6}\alpha_a\alpha_b r \sin 2\theta \frac{\cos 2\theta}{r^3}. \tag{37}$$

It is obvious that this model is similar to Kuhn's classical two-oscillator model (see Fig. 17).

Let us now derive Equation (37) directly. The electric field of a wave directed along the x axis forces both particles to oscillate in phase; the dipole

$$p_x = \alpha(E\cos\theta - Tp_x). \tag{38}$$

is induced in each particle. If, for purposes of simplicity, we limit ourselves to the single frequency ω_0 in the particle spectrum, the polarizability of the particle [see Equation (26a)] is

$$\alpha = \frac{e^2}{m}\frac{f_0}{\omega_0^2 - \omega^2}, \tag{39}$$

where ω_0 and ω are the angular frequencies. Substituting the value of T from Equation (36) and α into Equation (38), we find

$$p_x = \frac{e^2}{m}\frac{f_0\cos\theta}{\omega_0^2 + \frac{e^2}{m}f_0\frac{\cos 2\theta}{r^3} - \omega^2}E = \alpha_x E; \tag{40}$$

the resultant dipole moment of the dimer is then

$$P_x = 2p_x\cos\theta. \tag{41}$$

Comparison of α_x with α shows that the dipole—dipole interaction alters the transition frequency in the following manner:

$$\omega_x^2 = \omega_0^2 + \frac{e^2}{m}f_0\frac{\cos 2\theta}{r^3}. \tag{42}$$

A field directed along the y axis causes the particles to oscillate with opposite phases. In this case,

$$p_y = \alpha (E \sin \theta + T p_y), \qquad (38a)$$

$$p_y = \frac{e^2}{m} \frac{f_0 \sin \theta}{\omega_0^2 - \frac{e^2}{m} f_0 \frac{\cos 2\theta}{r^3} - \omega^2} E = \alpha_y E, \qquad (40a)$$

$$P_y = 2 p_y \sin \theta, \qquad (41a)$$

$$\omega_y^2 = \omega_0^2 - \frac{e^2}{m} f_0 \frac{\cos 2\theta}{r^3}. \qquad (42a)$$

Instead of the single frequency ω_0, we have thus obtained two frequencies, ω_x and ω_y, with a separation

$$\omega_x - \omega_y \approx \frac{\omega_x^2 - \omega_y^2}{2\omega_0} = \frac{e^2 f_0 \cos 2\theta}{m \omega_0 r^3}. \qquad (43)$$

Instead of a single transition with the frequency ω_0, transitions with the frequencies ω_x and ω_y, polarized in mutually perpendicular directions, have developed. The polarizabilities corresponding to these transitions are written in the form

$$\alpha_x = \frac{e^2}{m} \frac{f_0}{\omega_x^2 - \omega^2}, \quad \alpha_y = \frac{e^2}{m} \frac{f_0}{\omega_y^2 - \omega^2}, \qquad (44)$$

since, given our assumptions, the oscillator strengths f_0 remain constant.

Our model lacks a plane or center of symmetry. Under the action of a field, it therefore acquires the magnetic moment

$$M' = \frac{1}{2c} [r \dot{P}] = \frac{1}{c} \beta \dot{E}, \qquad (45)$$

translation along the x axis corresponds to the magnetic moment (directed along the y axis)

$$M'_{(x)} = -\frac{1}{2c} r \sin \theta \, p_x = -\frac{\alpha_x}{4c} r \sin 2\theta \, \dot{E}_x = \frac{1}{c} \beta_x \dot{E}_x, \qquad (46)$$

while translation along the y axis corresponds to the moment

$$M'_{(y)} = \frac{1}{2c} r \cos \theta \, p_y = \frac{\alpha_y}{4c} r \sin 2\theta \, \dot{E}_y = \frac{1}{c} \beta_y \dot{E}_y. \qquad (46a)$$

The quantity β_4, which is defined by Equation (30), has the sense of the average of β_x, β_y, and β_z. We find from Equations (46) and (46a) that

$$\beta_x = -\frac{\alpha_x}{4} r \sin 2\theta, \quad \beta_y = \frac{\alpha_y}{4} r \sin 2\theta \tag{47}$$

and

$$\beta = \frac{1}{3}(\beta_x + \beta_y) = \frac{1}{12} r \sin 2\theta \, (\alpha_y - \alpha_x)$$

$$= \frac{1}{12} r \sin 2\theta \, \frac{e^2}{m} f_0 \left(\frac{1}{\omega_y^2 - \omega^2} - \frac{1}{\omega_x^2 - \omega^2} \right)$$

$$= \frac{1}{12} r \sin 2\theta \, \frac{e^2}{m} f_0 \, (\omega_x^2 - \omega_y^2) \; \frac{1}{(\omega_y^2 - \omega^2)(\omega_x^2 - \omega^2)} . \tag{48}$$

Hence, according to Equation (43),

$$\beta_4 = \frac{1}{6} r \sin 2\theta \, \frac{e^2}{m} f_0 \, \frac{e^2}{m} f_0 \, \frac{\cos 2\theta}{r^3} \, \frac{1}{(\omega_y^2 - \omega^2)(\omega_x^2 - \omega^2)}$$

$$\approx \frac{1}{6} \alpha_a \alpha_b r \sin 2\theta \, \frac{\cos 2\theta}{r^3} \sim \frac{1}{(\omega_0^2 - \omega^2)^2} . \tag{49}$$

We have thus obtained Equation (37).

The contribution made by polarization to the optical activity is thus expressed by the product of the polarizabilities. The frequency ω_0 in the dispersion formula (49) and its analogs corresponds to the electric dipole transitions represented in the expression for polarizability (26a). In other words, these are strong shifts in absorption, to which relatively high values of p_{0j}, i.e., f_j, correspond. Hence it follows that the contribution made by β_4 to β is not fully represented by this quantity. In actuality, the rotatory strength (28) is expressed by the product of p_{0j} and μ_{j0}. High values of p_{0j} usually correspond to low values of μ_{j0}, and vice versa. The optical activity can therefore undergo weak transitions with small p_{0j}, which do not make any marked contribution to the absorption and thus to the polarizability or β_4. The contribution made by weak transitions is taken into account by the term β_3 (see p. 62).

In calculating β_4, we made assumptions that are not always valid. Representation of the dipole—dipole interaction by Equation (31) assumes that the distance between the dipoles materially exceeds their dimensions. This may not be the case within a molecule. Calculations made from the theory of polarizability yield more reliable results with respect to the influence of intermolecular interaction on optical activity [115].

The value of β_3 is calculated with the aid of the so-called single-electron model [116, 117]. Ketones have been studied in particular detail, considering the behavior of the electron of the chromophoric $C = O$ group, which is responsible for the longest-wavelength absorption of a molecule in the asymmetric field of the remaining atoms.

The relationship of the different contributions made to β is analyzed by Kruchek [118], while a semiempirical computation technique is suggested by Vol'kenshtein and Levitan [119, 120].

This is the current state of the theory. Until recently, little use was made of data on optical activity for studying molecular structure, despite its high sensitivity to this factor. Ordinary measurements of specific rotation at a single wavelength (the sodium D line has been most frequently employed) did not yield any significant information. The complex theory of the phenomenon did not provide any practical opportunity for interpretation of the data obtained. The situation subsequently changed when it was found that study of optical rotatory dispersion (ORD) can make it possible to reach very important conclusions and to study details of molecular structure inaccessible to other research techniques. In particular, ORD permitted determination of the conformations of very complex molecules, such as terpenes and steroids [121]. Correspondence between experimental results and theoretical calculations was obtained in investigating ORD rather than optical activity at a single wavelength.

Study of the anomalous optical rotatory dispersion (AORD) in the region of the natural absorption bands of a compound yields especially valuable information. The formulas for polarizability α and the optical-activity parameter β [see Equations (24) and (26)] do not take absorption into account and therefore give an infinitely large result at $\nu = \nu_j$. Equation (26) must be replaced by the equation [63]

$$\alpha = \frac{2}{3h} \sum_j \frac{\nu_j D_{0j}}{\nu_j^2 - \nu^2 + i\nu\Gamma_j} ,\qquad (50)$$

near the absorption bands; here, $D_{0j} = p_{0j}^2$ is the dipole strength, Γ_j is the damping parameter (with the sense of the band half-width on the frequency scale), and $i = \sqrt{-1}$. In similar fashion [63, 117],

$$\beta = \frac{c}{3\pi h} \sum_j \frac{R_{0j}}{v_j^2 - v^2 + iv\Gamma_j}. \tag{51}$$

The complex expressions for α and β indicate that the refractive index and the rotation of the plane of polarization are complex, i.e.,

$$\tilde{n} = n - i\varkappa, \tag{52}$$

$$\tilde{\varphi} = \varphi - i\theta. \tag{53}$$

The quantity $\varkappa$ in Equation (52) represents the absorption index. If light with an intensity I_0 strikes a layer of a given compound with thickness l, the intensity of the transmitted light is

$$I = I_0 \exp\left[-\frac{4\pi\varkappa}{\lambda} l\right], \tag{54}$$

or, introducing the absorption coefficient $k = 4\pi\varkappa/\lambda$,

$$I = I_0 e^{-kl}. \tag{54a}$$

The value of D_{0j} is found experimentally from the dependence of $\varkappa$ or k on frequency:

$$D_{0j} = \frac{3hc}{8\pi^3 N_1} \int_0^\infty \frac{k_j(v)}{v} dv.$$

The polarizability α, and hence the refractive index n, are thus inseparably associated with absorption. If a compound did not absorb light at some wavelengths, all the D_{0j} would be zero and its refractive index would be one. The refraction of a transparent substance represents absorption in another spectral region.

The imaginary portion of the factor $\tilde{\varphi}$, which is designated by θ, characterizes not merely the absorption, but the difference in the absorption of waves circularly polarized to the right and left, i.e., the circular dichroism [compare Equation (19)]:

$$\tilde{\varphi} = \frac{\pi}{\lambda} (\tilde{n}_l - \tilde{n}_r) l = \frac{\pi}{\lambda} (n_l - n_r) l - i \frac{\pi}{\lambda} (\varkappa_l - \varkappa_r) l. \tag{56}$$

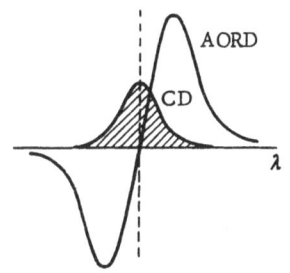

Fig. 19. Anomalous optical rotatory dispersion (AORD) and circular dichroism (CD).

As a result of this nonuniform absorption of right- and left-handed waves, an optically active compound not only rotates the plane of polarization in an absorption band, but converts linearly polarized light to elliptically polarized light. The ellipticity is measured again by the parameter θ, which equals

$$\theta = \frac{\pi}{\lambda}\,(\varkappa_l - \varkappa_r)\,l. \qquad (57)$$

The rotatory strenth is expressed by means of θ in the same fashion as the dipolar strength is expressed by means of k, i.e.,

$$R_{0j} = \frac{3hc}{8\pi^3 N_1} \int\limits_0^\infty \frac{\theta_j\,(\nu)}{\nu}\,d\nu. \qquad (58)$$

Optical activity is related to circular dichroism in the same manner as refraction is related to absorption. Figure 19 is a graph of the wavelength dependence of AORD and circular dichroism (CD). It is customary to refer to AORD as the Cotton effect in the current literature. Actually, Cotton discovered circular dichroism rather than AORD. These quantities are inseparably associated but not identical.

Measurement of the AORD or CD permits direct determination of an important constant, the rotatory strength. The AORD and CD are interrelated and, using the so-called Kronig—Kramers transforms, one set of data can be recalculated for the other factor [122]. Analysis of appropriate measurements has shown that the sensitivity (signal-to-noise ratio) is higher for CD measurements than for AORD measurements [123]. There are cases in which no AORD is observed but the CD is noticeable.

Experimental techniques for measuring ORD, AORD, and CD have improved markedly in recent years. There are commercial spectropolarimeters and dichrographs suitable for work in the regions of absorption bands extending into the far ultraviolet (wavelengths of less than 200 mμ), which is especially important

in studying proteins. The principal difficulty with which one is
confronted in this case is determined by actual absorption.
In order to study the rotation of the plane of polarization with a
spectropolarimeter, it is necessary to use either a thin layer or a
low-concentration solution of the compound in question. The absolute values of φ and θ are very low and high instrument sensitivity (up to 0.001° or even 0.0001°) is required.

The broad experimental and theoretical research that has
been conducted on optical activity has undoubtedly been stimulated
by biology, particularly since spectropolarimetry is a powerful
tool for studying the structure of biopolymers.

Appendix II discusses another spectropolarimetric method,
based on magnetic rotation of the plane of polarization.

Chapter 8

OPTICAL ACTIVITY
OF POLYPEPTIDE CHAINS

Proteins are asymmetric molecules and consequently have optical activity. The asymmetry of proteins ultimately results from that of carbohydrates. A protein is synthesized on template RNA through the intermediary of tRNA molecules. The asymmetry of RNA is caused by that of the carbohydrate groups of ribose. It is precisely this factor that makes *d*- and *l*-amino acids nonequivalent with respect to incorporation into polypeptide chains. Proteins are built up from *l*-amino acids [3]. We should be grateful to nature for this, since optical activity gives us valuable opportunities for studying their structure.

Substantial changes in optical activity and ORD occur during denaturation of a protein. Optical activity is naturally greatly altered during cooperative helix—random coil transitions. In the helical state, since the helix as a whole is asymmetric, the measured optical activity can roughly be considered to consist of two components: the optical activity of the helix and that of the individual amino acid residues (except the symmetric Gly residues). Only the second component occurs in a randomly folded structure [3, 7, 8].

The ORD of the α-helix differs materially from that of a disordered random coil. Study of the ORD of a protein makes it possible to determine the degree of helicity and the changes in this factor during conformational transformations.

The AORD and CD at ultraviolet absorption bands of a protein provide very similar information on protein-chain conformations.

71

In some cases, symmetric molecules attached to native proteins and polyamino acids become optically active and exhibit
AORD in their natural absorption bands. This effect usually disappears after denaturation. If such molecules have absorption bands
in the near-ultraviolet (coenzymes) or visible (dyes or heme) spectral regions, study of the effect provides information on protein
conformations without the need to resort to difficult measurements
in the far-ultraviolet region.

The ORD of most optically active compounds is expressed by
the one-term empirical equation

$$[M] = a_0 \frac{\lambda_0^2}{\lambda^2 - \lambda_0^2} = \frac{K}{\lambda^2 - \lambda_0^2}. \tag{59}$$

An approximation of this equation is obtained from Equations (23)
and (24), assuming that the sum of dispersions in a region remote
from the area of natural absorption can be replaced by a single
term with the effective frequency ν_0 or the wavelength $\gamma_0 = c/\nu_0$;
a_0 and $K = a_0\lambda_0^2$ are constants and λ is the wavelength of the incident light. Equation (59) is called the Drude equation, since Drude
derived it with the aid of an extremely simplified classical molecular model, using a single electron moving in a helical path. The
Drude equation provides a satisfactory description of the ORD of
polyamino acids in the randomly folded state. In this case, λ_0 is
268 mμ. On the other hand, the ORD of α-helical polyamino acids
is substantially better described by the Moffitt equation [3, 124, 125]

$$[M] = a_h \frac{\lambda_0^2}{\lambda^2 - \lambda_0^2} + b_0 \frac{\lambda_0^4}{(\lambda^2 - \lambda_0^2)^2}. \tag{60}$$

The rotation for a protein containing both α-helical and disordered
segments is made up of Equations (59) and (60). If we designate
the degree of helicity as f,

$$[M] = a_0 \frac{\lambda_0^2}{\lambda^2 - \lambda_0^2} + f\left(a_h \frac{\lambda_0^2}{\lambda^2 - \lambda_0^2} + b_0 \frac{\lambda_0^4}{(\lambda^2 - \lambda_0^2)^2}\right). \tag{61}$$

The value of f can be found from the dependence of [M] on
λ, i.e., from the ORD. For example, if we plot the quantity [M]
$(\lambda^2/\lambda_0^2 - 1)$ along the ordinate and $(\lambda^2/\lambda_0^2 - 1)^{-1}$ along the abscissa,
the slope of the resultant line is fb_0 and the segment cut off on the
ordinate is $a_0 = fa_h$. The constants a_0, b_0, and a_h are found em-
pirically from data on denatured and native α-helical polyamino

acids. The following values are generally employed: $\lambda_0 = 212$ mμ, $a_h = 650$, and $b_0 = -630$ deg $\cdot$ cm^2 $\cdot$ decimole^{-1}. A minus sign on b_0 corresponds to a right-handed α-helix; this quantity is positive for a left-handed helix. Other values have been proposed, including $\lambda_0 = 220$ mμ and $b_0 = -390$ (a modified version of the Moffitt equation) [126]. Determination of the degree of α-helicity from ORD with the aid of Equation (61) or by other methods (see p. 79) yields empirical values for f. It is important to note that, in the few cases where f has been determined directly by x-ray diffraction analysis (for hemoglobin and myoglobin), spectropolarimetry yielded similar values for this factor. The values of f given in Table 3 (see p. 11) were obtained by the ORD method.

Does Moffitt's equation have any basis other than an empirical one?

For a spectral region remote from the area of natural absorption, one can always represent [M] in the form [compare Equations (23) and (24)]

$$[M] = \sum_j \frac{a_j \lambda_j^2}{\lambda^2 - \lambda_j^2} . \tag{62}$$

Expanding this expression into a Taylor series for $\lambda_j^2 - \lambda_0^2$ (λ_j^2, $\lambda_0^2 \ll \lambda^2$), we obtain [127]

$$[M] = \sum_j \frac{a_j \lambda_0^2}{\lambda^2 - \lambda_0^2} + \sum_j \frac{a_j \lambda^2 (\lambda_j^2 - \lambda_0^2)}{(\lambda^2 - \lambda_0^2)^2} + 0\left(\frac{1}{(\lambda^2 - \lambda_0^2)^3}\right). \tag{63}$$

Transforming the first term of this expression:

$$\frac{\lambda^2 (\lambda_j^2 - \lambda^2)}{(\lambda^2 - \lambda_0^2)^2} \equiv \frac{(\lambda^2 - \lambda_0^2 + \lambda_0^2)(\lambda_j^2 - \lambda_0^2)}{(\lambda^2 - \lambda_0^2)^2} \equiv \frac{\lambda_j^2}{\lambda^2 - \lambda_0^2} - \frac{\lambda_0^2}{\lambda^2 - \lambda_0^2} + \frac{\lambda_0^2 (\lambda_j^2 - \lambda_0^2)}{(\lambda^2 - \lambda_0^2)^2},$$

summing the first and second terms, we find

$$[M] = \sum_j \frac{a_j \lambda_j^2}{\lambda^2 - \lambda_0^2} + \sum_j \frac{a_j \lambda_0^2 (\lambda_j^2 - \lambda_0^2)}{(\lambda^2 - \lambda_0^2)^2} + 0\left(\frac{1}{(\lambda^2 - \lambda_0^2)^3}\right). \tag{63a}$$

Introducing the notation

$$a_0 = \sum_j a_j \frac{\lambda_j^2}{\lambda_0^2}, \quad b_0 = \sum_j a_j \left(\frac{\lambda_j^2}{\lambda_0^2} - 1\right),$$

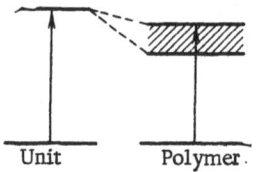

Fig. 20. Diagram of exciton splitting of spectral levels in regular system.

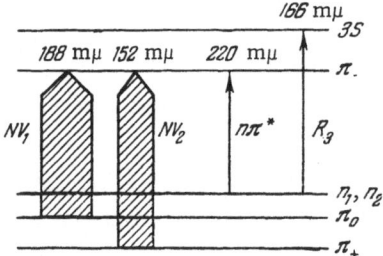

Fig. 21. Diagram of energy levels of amide group.

we obtain

$$[M] \approx \frac{a_0\lambda_0^2}{\lambda^2 - \lambda_0^2} + \frac{b_0\lambda_0^4}{(\lambda^2 - \lambda_0^2)^2}. \qquad (63b)$$

Equation (63b) is fully consistent with Equation (60). The sign on b_0 can actually be the opposite of that on a_0 if $\lambda_0 > \lambda_j$. We have thus demonstrated that Moffitt's equation is possible, but its validity for the α-helix has not been proved.

Moffitt derived Equation (60) theoretically. In his first article [128], he proceeded from the theory of the exciton spectrum of a regular α-helical polypeptide, which is based on the theory of molecular-crystal spectra developed by Davydov [129]. If there is a regular aggregate of chromophoric groups (in proteins, these are the peptide groups —CO—NH—), resonance energy transfer can occur between their excited energy levels. Propagation of an excitation wave (an exciton) is therefore possible in the regular system. The interaction causes splitting of the energy levels to form a broad zone, as is shown diagrammatically in Fig. 20. The selection rules permit transitions only in strictly defined levels of the zone during absorption and emission of light; band polarization is directly related to molecular-crystal symmetry [107, 129].

An individual peptide group is similar to an amide molecule R_1—CO—NH—R_2. Study of the electron spectra of simple amides has established the pattern of energy levels and transitions shown in Fig. 21 [130, 131]. The width of the arrow symbolizes the dipolar strength of the corresponding transition. Figure 22 shows the wave functions of an amide corresponding to its most labile electron. The π_+ level in Fig. 21 corresponds to the bonding orbital of the CO, while the π_- level corresponds to the nonbonding orbital, i.e.,

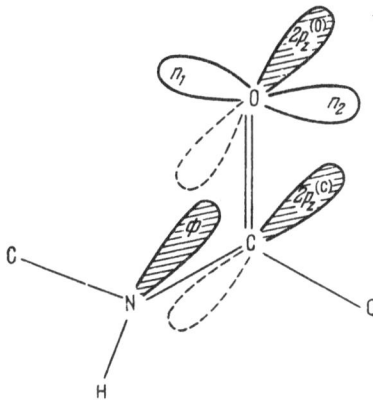

Fig. 22. Wave functions of amide group.

$$\pi_+ = (2p_z^{(0)} + 2p_z^{(c)})/\sqrt{2}, \quad \pi_- = (2p_z^{(0)} - 2p_z^{(c)})/\sqrt{2}.$$

The π_0 level corresponds to the nonbonding orbital of the nitrogen, n_1 and n_2 are the states of the unshared electron pair of the oxygen, and R_3 corresponds to Rydberg's atomic transition in the oxygen. It is therefore understandable that the $n\pi$-transition has a low dipolar strength: the n- and π-electron clouds are almost perpendicular to one another and overlap only slightly.

Taking into account the dipole—dipole interaction of the peptide groups of the α-helix, Moffitt calculated the exciton splitting of the $\pi_0\pi$—band near 190 mμ. In accordance with Davydov's theory, this band should be split into two components, one of which is polarized along the axis of the helix, and the other perpendicular to it. According to Moffitt's estimate, the splitting $\Delta\nu = \nu_{\parallel} - \nu_{\perp}$ is 1500 cm^{-1}. This value is close to the experimentally observed splitting [132].

Quantum-mechanical calculation of the optical activity of the α-helix, based on the dipole—dipole polarization interaction, led Moffitt to the equation

$$[M] = \sum_j \left(\frac{A_j \nu^2}{\nu_j^2 - \nu^2} + \frac{B_j \nu^2 \nu_j \Delta\nu_j}{(\nu_j^2 - \nu^2)^2} \right), \tag{64}$$

which is similar to Equation (60). The idea underlying this derivation is made clearer by recalling the calculation of the polariza-

tion contribution to the optical activity of a dimer made above. If
we limit ourselves to considering the electrostatic interaction, the
exciton splitting and associated molecular optical properties can
be treated in a purely classical manner. For a dimer, we obtained
the formula (see p. 66)

$$\beta \sim \frac{1}{(\nu_0^2 - \nu^2)^2}.$$ (49)

The corresponding molecular rotation is written in the form

$$[M] \sim \frac{\nu^2}{(\nu_0^2 - \nu^2)^2},$$ (49a)

or, converting from frequencies to wavelengths,

$$[M] \sim \frac{\lambda^2 \lambda_0^4}{(\lambda^2 - \lambda_0^2)^2} \equiv \lambda_0^2 \left(\frac{\lambda_0}{\lambda^2 - \lambda_0^2} + \frac{\lambda_0^4}{(\lambda^2 - \lambda_0^2)^2} \right).$$ (49b)

The resultant equation contains a quadratic term, as does the Moffitt equation. However, both terms have the same sign, while experimentation has yielded values of a_h and b_0 for the α-helix that
are similar in numerical value but have opposite signs. The model
in question is undoubtedly overly primitive. Nevertheless, it is
adequate to enable us to understand why exciton splitting of energy
levels occurs and how the quadratic term in the dispersion equation arises.

An article by Moffitt, Fitts, and Kirkwood [133] gave a new
calculation of the ORD of the α-helix on the basis of the theory of
polarizability which, as we have seen, yields a description of exciton splitting. It was demonstrated that an error associated with
incorrect evaluation of the boundary conditions of the problem had
crept into Moffitt's calculation of the ORD [128] (but not into his
calculation of the splitting). Moffitt, Fitts, and Kirkwood [133] obtained a general theoretical equation similar to Moffitt's equation
and, in a later article [133a], attempted a quantitative evaluation
of the constants and compared the theoretical and experimental
data; the results obtained, however, were ambiguous. The theory
of polarizability also figured in the work of Murakami [134], whose
results were almost identical to those given by Moffitt, Fitts, and
Kirkwood [133].

In a number of later articles, Tinoco developed a theory of the optical properties of polymers based on the same notions arising from the exciton splitting of monomer energy levels, which results from the dipole—dipole interaction. This author devised a theory to account for the absorption spectrum [135-140] and ORD [136, 139-143] of biopolymers, specifically applying it to double helical polynucleotides and α-helical polypeptide chains. Several of his articles [132, 144, 145] contain a general review of his work.

The principal finding of these investigations with respect to the problem in which we are interested (the ORD of the α-helix) was that the expression for the ORD is more complex than Moffitt's equation, containing additional terms. However, a quadratic term always arises in an ordered system; the coefficient of this term depends on the character of the intergroup interaction.

McLachlan and Ball [114] gave a simplified version of the theory, employing the quantum-mechanical method of a self-consistent time-dependent field. It was demonstrated above (see p. 64), using the two-oscillator model, that the same results can be reached on the basis of the classical theory of polarizability. These authors also considered two particles to explain the hypochromic effect, i.e., the fact that the long-wave absorption band of a peptide group has lower intensity in the α-helix than in a random coil, but they assumed that these particles had a whole set of frequencies rather than the single frequency ω_0. If the rod-like particles of the dimer are parallel, the intensity of the lowest-frequency transition is reduced as a result of the increase in the intensities of the other transitions; the resultant oscillator strength is

$$F_j = 2f_j - \frac{2e^2 T}{m} {\sum_k}' \frac{2f_j f_k}{\omega_j^2 - \omega_k^2} , \qquad (65)$$

while the total oscillator strength of the two independent particles is $2f_j$. Equation (65) is similar to that obtained by Tinoco and his colleagues, but its derivation is simpler. We might mention the articles by Nesbet [146] and DeVoe [147], which are also devoted to the theory of hypochromism. In the second section of their article, McLachlan and Ball moved on from dimers to polymers and, specifically, to the α-helix. They used the same basis for a comparatively simple derivation of the ORD equations; the coefficients

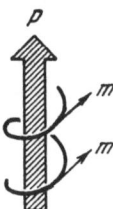

Fig. 23. Electric (p) and magnetic (m) moments of n π-transition in α-helix.

in these equations were related to the geometric parameters of the helix. However, the calculations made by these authors were not reduced to numerical results. The theory devised by McLachlan and Ball was generalized and expanded in a recent article by Harris [148]. Let us now turn to the quantitative aspect of the problem. As has already been mentioned, Moffitt considered only the $\pi\pi$-transitions in the peptide bond. This transition actually corresponds to the most intense band in the electronic spectrum. The theory based on the dipole—dipole interaction deals only with strong bands (the rotatory strength is expressed by the oscillator strength) and Moffitt was therefore consistent. Tinoco's specific calculations were also for the $\pi\pi$-transition.

However, it follows from the general equation for the rotatory strength [Equation (28)] that it can also be large for weak bands, provided that substantial matrix elements of the magnetic moment correspond to them. The contribution of these bands to the optical activity cannot be determined within the framework of the theory of polarizability, but the single-electron model permits a direct evaluation.

Proceeding from the single-electron model, Schellmann and Oriel [149] showed that a substantial matrix element of the magnetic moment corresponds to the nπ-band of a peptide group if the peptide groups of the molecule form a helix. The magnetic moments are tangential to the α-helix, while the electric moments are parallel to its axis (Fig. 23). These authors did not limit themselves to theoretical calculations, but also investigated the AORD in the natural-absorption region of a polypeptide near 200 mμ. They were able to demonstrate that the nπ-transition actually makes a noticeable contribution to the optical activity, although materially less than that of the $\pi\pi$-transition. It must be again emphasized that investigation of the AORD in an absorption band and of the CD permits quantitative determination of the rotatory strength. Figure 24 shows the trend of the AORD in the α-helix of polyglutamic acid and in its denatured, randomly folded form [149].

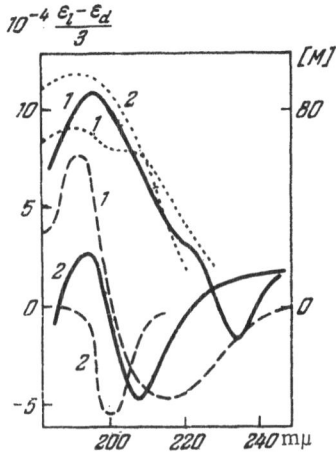

Fig. 24. Curves representing AORD
(solid line), CD (dash line), and absorption (dotted line) for polyglutamic acid. 1) Helix; 2) random coil.

The $\pi\pi$-band in the absorption spectrum of the α-helix is
split into two components, one polarized parallel ($\parallel$) to the helix
axis and the other polarized per-
pendicularly ($\perp$) to it. Tinoco
showed that, in calculating the optical activity, it is necessary to
take into account three rather
than two components. The perpendicular band is in turn split into
two bands. A curve involving the
superposition of the Cotton effects
should accordingly be obtained for
the AORD [143].

Table 10 compares the ex-
perimental and theoretical values
for the wavelengths, oscillatory
strengths, and rotary powers of
the α-helix (polyglutamic acid) [150]. The circular dichroism
(Cotton effect) was used to measure R [150]. It can be seen that
we can as yet speak only of semiquantitative agreement between
the theoretical and experimental data. The signs of R, the sequence of values, and the orders of magnitude are the same. It also follows from Table 10 that Moffitt's equation can be substanti-
ated only in general terms and not quantitatively; it must therefore
be regarded as semi-empirical. Nevertheless, it is quite reasonable to use this equation to determine the α-helicity of a protein,
since it yields results in agreement with the data given by other
methods.

There are also other empirical methods for determining the
α-helicity of proteins. Drude's equation

$$[M] = \frac{K_c}{\lambda^2 - \lambda_c^2}, \qquad (59)$$

is often employed; the value of λ_c depends on the helicity f. At a
helicity of 40%, $\lambda_c = 268\,m\mu$, while for a random coil, $\lambda_c = 212\ m\mu$.
Assuming the dependence of λ_c on f to be linear, the latter can be
determined from the ORD.

Table 10. Optical Properties of α-Helix

Transi-tion	Character of transition	Absorption			
		$\lambda, m\mu$		Oscillatory strength	
		theor.	exptl.	theor.	exptl.
$n_1\pi^-$		210—230	222	0.0001	0.007
$\pi^0\pi^-$	‖ parallel, in absorption spectrum	198	206	0.09	0.03
$\pi^0\pi^-$	⊥ perpendicular in absorption spectrum	188	189	0.16	0.10
$\pi^0\pi^-$	Considered only in rotation				

Transi-tion	Character of transition	Rotation			
		$\lambda, m\mu$		Rotary power, $R\cdot10^{40}$,erg.cm^3,rad	
		theor.	exptl.	theor.	exptl.
$n_1\pi^-$		210—230	222	—3.4	—22
$\pi^0\pi^-$	‖ parallel, in absorption spectrum	198	206	—126	—29
$\pi^0\pi^-$	⊥ perpendicular in absorption spectrum	191	190	+242	+81
$\pi^0\pi^-$	Considered only in rotation	185		—115	—

Blout and Shechter suggested that a two-term rather than one-term Drude equation be used for determining f [151, 152]. Instead of the effective value $\lambda_0 = 212$ or 220 mμ, this equation employs the experimentally observed absorption maxima $\lambda_1 = 193$ mμ and $\lambda_2 = 225$ mμ. The two-term Drude equation has the form

$$[\alpha] = \frac{A_1\lambda_1^2}{\lambda^2 - \lambda_1^2} + \frac{A_2\lambda_2^2}{\lambda^2 - \lambda_2^2}, \qquad (66)$$

with $A_1 > 0$ and $A_2 < 0$. These constants are linked by a linear relationship. In aqueous solutions of polyamino acids (with a dielectric constant $\varepsilon > 30$), we obtain

$$A_2 = -0.55 A_1 - 430,$$

while in organic solvents ($\varepsilon < 30$),

$$A_2 = -0.55 A_1 - 280.$$

The environment of the α-helix, i.e., the effective dielectric constant, can therefore be determined from the trend of the line A_2 (A_1).

For a randomly folded structure (1,1-copoly-l-methionine-l-methyl-S-cysteine in a 1 : 1 mixture of dichloroacetic and trifluoroacetic acids), $A_2 = 0$ and $A_1 = -600$. For an α-helix (poly-l-methionine in chloroform), $A_1 = 3020$ and $A_2 = -1900$. Hence, the helicity is

$$f_{193} = \frac{A_1 + 600}{36.2}, \quad f_{225} = -\frac{A_2}{19.0},$$

or

$$f = \frac{A_1 - A_2 + 600}{55.8}.$$

The results obtained with the one-term and two-term Drude equations are in general agreement with those obtained with Moffitt's equation. It is wise to process the ORD data by all available methods in order to obtain mutual checks.

As was pointed out in Chapter 2, the α-helix is not the only ordered conformation for polypeptide chains.

It is obvious that the β-form (see p. 11) should be characterized by a specific ORD only if it is asymmetric as a whole. Such asymmetry apparently does exist: the β-form is not completely planar, but is folded in an accordion-like manner (Fig. 25). Imahori suggested an expression similar to Moffitt's equation, but with different constants for the ORD of the β-form [153]. Wada and his colleagues determined these constants experimentally for a single compound [154]. The value of b_0 proved to be very small. Recent theoretical calculations have also shown that b_0 is close to zero [155].

Very little research has been done on the optical activity of polypeptide β-forms and other ordered conformations differing from the α-helix.*

---

*A number of articles devoted to spectropolarimetric investigation of the β-forms of proteins and polyamino acids have recently appeared: Sarkar and Doty, Proc. Nat. Acad. Sci. USA, 55: 981 (1966); Iisuka and Yang, Proc. Nat. Acad. Sci. USA, 55: 1175 (1966); Townend et al., Biochem. Biophys. Res. Comm., 23: 163 (1966);

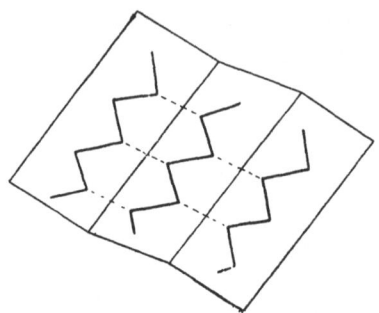

Fig. 25. β-Form of polypeptide.

Throughout this discussion, we have been speaking of the average scalar value of the optical activity. It is with this factor that one must deal in studying protein and polypeptide solutions. However, the optical-activity parameter β is a tensor quantity, having different values for different intramolecular directions. If the molecule were oriented, it would be possible to obtain more information by measuring the optical activity along and across the α-helix or other anisotropically ordered structure. Unfortunately, it is difficult to orient proteins and polypeptides with an electric field; being polyelectrolytes, they are subject to electrophoresis and it is therefore necessary to make measurements under a pulsed regime. Here we encounter the complex relaxation behavior of macromolecules in an ionic environment. The current state of the theory does not permit interpretation of results obtained under such conditions in terms of biopolymer structure. Measurements made with a nonelectrolyte (poly-γ-benzylglutamate) oriented in an electric field actually showed its optical activity to be anisotropic [145]. On the whole, however, no research at all has been done on the anisotropy of optical activity in proteins and polyamino acids.

Can the change in optical activity during partial or complete denaturation of a protein be attributed solely to disappearance of the helical segments? The tacitly assumed affirmative answer to this question cannot be regarded as sufficiently well-grounded. When a globule unfolds, the environment of the asymmetric amino acid residues is altered and the optical activity should change merely as a result of the change in intermolecular interaction. This approach has a solid physical basis, but no appropriate experimental or theoretical investigations have been conducted. It would obviously be interesting to make a detailed study of the change in optical activity during denaturation of globular proteins with a low α-helicity, such as ribonuclease.

Davidson et al., Biochem. Biophys. Res. Comm., 23: 156 (1966); Timasheff, Townend, and Mescanti, J. Biol. Chem., 241: 1863 (1966); Townend, Kumosinski, and Timasheff, J. Biol. Chem., 242: 4538 (1967).

As for induced optical activity, which was mentioned above, it will be considered below.

Chapter 9

OPTICAL ACTIVITY AND CONFORMATIONAL PROPERTIES OF ENZYMES

We showed above that study of the ORD outside the absorption region and in the absorption bands themselves provides valuable information on the conformational properties of proteins. We also mentioned another phenomenon: the development of AORD in the absorption bands of small molecules of dyes and other compounds attached to proteins.

Both processes are of great importance for research on enzymes. Investigation of ORD is now one of the most important techniques for determining protein conformations and, consequently, changes in structure. The second phenomenon opens up two possibilities. First of all, one can introduce a tag into an enzyme molecule, i.e., attach a dye to the molecule and investigate its AORD in the visible region of the spectrum. As we have seen, this anomalous dispersion of optical activity should be sensitive to the conformation of the enzyme. Secondly, many enzymes are active only in the presence of a coenzyme or contain prosthetic groups. In most cases, the coenzymes or prosthetic groups are conjugated π-electron systems [156]; their absorption bands consequently fall into the near-ultraviolet and visible regions. Investigation of AORD is far easier than in the region of natural protein absorption. The AORD of a coenzyme or prosthetic group is also sensitive to the conformation of the apoenzyme and to chemical transformations in the active center.

Development of AORD in the absorption bands of dyes bonded to biopolymers was first detected by Blout and Stryer [157]. A dye molecule is symmetric and has no optical activity in the absence of a biopolymer. The AORD disappears when the biopolymer is denatured. Blout and Stryer considered three models for development of optical activity in a dye (compare [3]).

1. Asymmetry develops in the dye molecule under the action of an asymmetric biopolymer.

2. The dye molecules form a secondary helix around the polypeptide helix. The optical activity results from the structure of the secondary helix as a whole.

3. The dye molecules polymerize and form helices that extend alongside the polypeptide helix, producing a sort of grafted polymer.

Blout and Stryer consider the first model to be improbable, since the effect disappears after denaturation, although the dye remains bonded to the asymmetric amino acid residues. However, experimentation favors the first model. The effect is observed at a relative dye concentration (in acridine orange—protein systems) so low that we cannot speak of a secondary helix [159]. The third model has received no confirmation whatsoever. Theoretical calculations made by the methods of quantum chemistry (with the dye molecule simulated by a potential well) have shown that the effect apparently consists in induced optical activity [160]. The electron shell of the dye molecule becomes asymmetric under the influence of the helical biopolymer; this is impossible in the denatured state. The results obtained by Permogorov and Lazurkin in investigating the ORD of complexes of DNA and acridine dyes [161, 162] lead to a similar conclusion. It is interesting that these authors were able to observe asymmetric dye dimers in the complexes, for which the AORD curve was symmetric and described by Equation (49). The calculation made in Chapter 7 is applicable to such dimers.

Very little work has been done on the ORD and AORD of enzymes. However, there is no doubt that such investigations will become quite common. Li, Ulmer, and Vallee [163] used the AORD method to investigate the interaction of a coenzyme, the reduced form of nicotinamide adenine dinucleotide (NADH) (with an absorption band at 327 mμ), and its analogs with liver dehydrogenase. From the trend of the AORD in the absorption bands of de-amino-NADH, 3-acetylpyridine-NADH, and trinicotinamide-NADH, they were able to determine the stoichiometric relationships in the complex and to investigate its dissociation under the action of n-chloromercuribenzoate. This work is principally of methodological interest.

Breusov and his colleagues, working in Braunshtein's laboratory, studied the circular dichroism in the absorption band of the coenzyme pyridoxal phosphate in aspartate transaminase and showed that structural investigations of enzymes can be based on the existence of the aforementioned effect [164]. Torchinskii and Koreneva used induced AORD in the absorption bands of the coenzyme to conduct systematic investigations of aspartate—glutamate transaminase in the same laboratory. The trend of the AORD made it possible to detect structurally different forms of the enzyme and its intermediate complexes, and to determine the details of the chemical mechanism of the corresponding enzymatic reactions. The AORD proved to be very sensitive to interactions between the enzyme and other molecules. It disappeared when NH_2OH or NH_2-NH_2 was added; it persisted but changed sign when the semicarbazide, thiosemicarbazide, or hydrazide of isonicotinic acid was added. Inhibition of the enzyme directly affected the AORD. The results obtained were convincingly interpreted on the basis of changes in the chemical bonds [165, 166].

Aki, Takagi, Isémura, and Yamano studied the AORD and CD of d-amino acid oxidase and established that a change in the state of the coenzyme has no effect on the helicity of the apoenzyme [167]. However, a_0 is markedly altered when the coenzyme is split off. Hence, it was concluded that there are conformational changes in the nonhelical portion of the protein, which are accompanied by a marked decrease in the denaturation resistance of the apoenzyme in comparison with the holoenzyme [167]. A number of articles have been devoted to heme-containing proteins: catalase, cytochromes, hemoglobin, and myoglobin. These will be discussed in Chapter 13.

Let us now turn to the conformational transformations of enzymes during their interactions with substrates, coenzymes, and other effectors.

Platt and Niemann [168] studied the hydrolysis of acetyl-l-leucylmethyl-α-chymotrypsin at different medium pH's. The

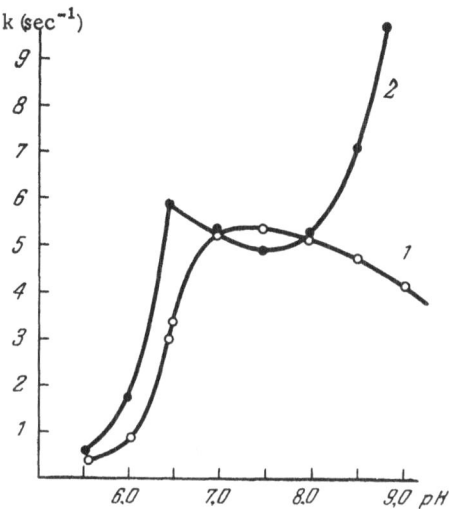

Fig. 26. Enzymatic activity of α-chymotrypsin without effector (1) and in presence of benzamide (2) (after Platt and Niemann).

variation in enzymatic activity described a bell-shaped curve with a maximum at pH ≈ 7.5 (see Chapter 10). The enzyme activity was altered in the presence of benzamide. This compound acted as an activator at low pH's, as an inhibitor at pH 7-8, and again as an activator at pH > 8 (Fig. 26). The authors believe that these results were produced by changes in the conformation of the active center as the pH was varied. They surmise that the activation of the enzyme by benzamide was due to formation of ternary enzyme—substrate—activator complexes having an increased reactivity. In view of the fact that the substrate and activator are very weakly ionized over the pH range investigated, it can be assumed that the changes observed resulted from changes in the enzyme conformation. This important hypothesis should be verified by independent means.

Havsteen and Hess [169] noted small changes in the specific rotation of α-chymotrypsin during its interaction with substrates (diisopropylphosphofluoridate and n-nitrophenylacetate) and qualitatively attributed them to conformational changes. Parker and Lumry [170] detected changes in the optical activity of this enzyme when it was exposed to certain inhibitors and investigated the dependence of these changes on pH. Labouesse et al. [171] studied the specific rotation of acetylated chymotrypsin and its complex with diisopropylphosphofluoridate and found that the character of the dependence of rotation differed in these two cases. All the articles mentioned above and certain others provide information on conformational changes in an enzyme, but it is all indirect.

Bolotina et al. [172] used the ORD method to investigate the α-helicity of chymotrypsin in the presence and absence of benzamide as a function of medium pH. It was found that the dependence of this factor on pH is well correlated with that of enzymatic activity on pH, as established by Platt and Niemann. The results obtained are a direct proof that an enzyme undergoes conformational changes when the pH is varied and during its interaction with an activator.

Extremely interesting results have been obtained in spectropolarimetric investigations of dehydrogenases. These enzymes function in oxidation—reduction processes. Their action requires participation of the coenzyme nicotinamide adenine dinucleotide (NAD) or nicotinamide adenine dinucleotide phosphate (NADP) and their reduced forms (NADH and NADPH). The form involved depends on the direction of the reaction, i.e., on whether hydrogen is transferred from the substrate to the coenzyme or from the coenzyme to the substrate.

Lactic dehydrogenase (LDH) catalyzes the conversion of lactic acid to pyruvic acid, as well as the reverse reaction [173], i.e.,

$$LDH + NAD + l\text{-lactate} \rightleftharpoons LDH + NADH + pyruvate.$$

There are indirect data indicating changes in the conformation of LDH during its interaction with the coenzyme. The kinetics of deuterium exchange are altered in the presence of the coenzyme [174]. Attachment of the coenzyme and substrate (inhibitor) to LDH occurs in a definite sequence [175, 176]. Spectropolarimetry provides direct proof that the secondary structure of LDH changes during its interaction with NADH [177]. Table 11 presents data on the α-helicity of LDH, as determined from Equation (59). The α-helicity of LDH decreases by about 30% in the presence of NADH. The action of this enzyme is inhibited by a number of compounds. Oxalate is a competitive inhibitor with respect to lactate and a

Table 11. ORD Constants and Helicity of LDH and Its
Complexes

	λ_c, mμ	$-K_c \cdot 10^{-7}$	Degree of α-helicity, %
LDH in 0.1 M tris–HCl buffer at pH 7.0	277±5	0.95±0.055	46±3.5
LDH + NADH (1:5)	259±5	1.03±0.04	33±3.5
LDH + NADH + oxalate (2.5 · 10^{-2} M)	275±5	0.95±0.05	45±4.3
LDH + oxalate	276±3	0.95±0.05	46±2

Table 12. Influence of NAD and d,l-Glyceraldehyde on
ORD Constants of GAPD Apoenzyme (at pH 8.5)

System	λ_c, mμ	$K_c \cdot 10^{-6}$	a_0	b_0	A_{193}	A_{225}
Native GAPD + 2.5 moles NAD	268	96	—211	—200	409	—491
GAPD apoenzyme	252	114	—279	—131	245	—443
Reconstructed GAPD complex + 6 moles NAD	265	88	—150	—173	347	—429
GAPD + 2.5 moles NAD + d,l-GA	247	125	—209	—138	261	—439

noncompetitive inhibitor with respect to pyruvate. Kinetic and luminescence investigations have shown that a ternary enzyme—
coenzyme—inhibitor complex is formed [178].

As can be seen from Table 11, oxalate has a direct influence
on the conformational properties of LDH + NADH. Addition of
oxalate to this complex produced an α-helicity close to that for the
apoenzyme. Moreover, the α-helicity of the apoenzyme remains
unchanged when it is exposed to oxalate, perhaps because the inhibitor is not attached to the apoenzyme in the absence of the coenzyme [177].

The data obtained with the Drude equation are in agreement
with the results yielded by the Moffitt equation. Processing of the
ORD data by the method of Blout and Schechter (see p. 80) shows

that the α-helical segments of LDH are in a nonpolar environment corresponding to a dielectric constant $\varepsilon < 30$ [177].

D-glyceraldehyde-3-phosphate dehydrogenase (GAPD) catalyzes the oxidation of 3-phosphoglyceraldehyde (PGA) to 1,3-phosphoglyceric acid (PGA)

$$\text{GAPD} + \text{NAD} + \text{PGA} + \text{HPO}_4^{2-} \rightleftharpoons \text{GAPD} + \text{NADH} + \text{PGA}.$$

The reaction proceeds in several stages [172]. There are indirect proofs that the conformation of GAPD changes during its interaction with NAD. The coenzyme stabilizes GAPD with respect to proteolysis [180]. There are irreversible changes in viscosity and specific rotation when the coenzyme is split off [181]. The coenzyme probably serves as a conformational cofactor. Table 12 presents the results of a spectropolarimetric investigation of the conformational properties of GAPD [182, 183].

Reversible attachment of NAD to GAPD is possible: the enzyme crystallizes with 2.5 moles of bound NAD. It can be seen from Table 12 that, when the bound NAD is split off, there are reversible changes in the constant λ_c in the Drude equation, which correspond to a marked decrease in α-helicity. The changes in the constant b_0 in the Moffitt equation are in good agreement with these data.

D-glyceraldehyde is also a GAPD substrate, but the reaction proceeds far more slowly than with PGA. Addition of d,l-GA to GAPD with 2.5 moles of bound NAD causes formation of an enzyme-substrate complex that hydrolyzes during the second stage of the reaction. Since only 3 moles of GA and NAD participate in the transformations per mole of GAPD, they cannot produce any material changes in optical activity. Table 12 shows that the values of λ_c and K_c are materially altered after addition of d,l-GA, which can be attributed solely to conformational changes in GAPD.

NADH does not cause any substantial changes in the structure of GAPD. It can be surmised that the changes observed under the action of GA are due principally to conversion of NAD to NADH, which eliminates the influence of the NAD on the GAPD conformation.

Direct experiments have thus proved that the GAPD conformation is altered under the action of a substrate (GA) and a coenzyme (NAD but not NADH) [182, 183]. These data were fully confirmed by Listovsky et al. [184].

The α-helicity of GAPD decreases sharply under the action of 8 M urea: λ_c falls to 212-220 mμ and b_0 drops almost to zero [182]. Dodecyl sulfate (DDS) disrupts hydrophobic interactions and therefore breaks down the tertiary structure of proteins (see Chapter 3). Investigation of the influence of DDS on GAPD by hydrodynamic methods has shown that this compound causes a substantially greater change in the tertiary structure of GAPD than in that of LDH (compare [185]). When the dodecyl sulfate concentration is increased, LDH breaks down into subunits without any material change in their form. Analysis of the conformational properties of GAPD by the method of Blout and Schechter (see p. 80) shows that the effective dielectric permeability increases under the influence of DDS: the α-helix is surrounded by an aqueous rather than a hydrophobic environment [179]. Hence, it can be concluded that the GAPD structure is more labile than the LDH structure.

Tables 11 and 12, and other data, show that dehydrogenases contain a comparatively large number of α-helical segments [177, 183, 186, 187]. Such segments might induce AORD in attached symmetric chromophores (coenzymes). Moreover, as was demonstrated by Braunshtein, Torchinskii, and Koreneva [164-166], a coenzyme can acquire AORD as a result of a nonsymmetric interaction with adjacent groups rather than an interaction with an α-helical segment of the enzyme. This is confirmed by the existence of both positive and negative CD.

Alcohol dehydrogenase actually induces AORD in the 260 and 340 mμ absorption bands of NADH and in the 260 mμ bands of NAD [188]. Slight AORD has been detected in glutamate dehydrogenase, both in the bound NADH absorption band and in the vicinity of 290 mμ, i.e., in the vicinity of the natural absorption of the ordered aromatic amino acid residues of the enzyme. The GAPD–NAD complex is characterized by a broad absorption band at 340-400 mμ. No AORD is observed in this spectral region, but there is slight CD. This is not an artifact, since the CD disappears when GA is added, i.e., when the complex is broken up. Slight AORD has been detected in GAPD in the vicinity of 280 mμ [184]. No AORD has been observed in the 340 mμ absorption band of the complex with NADH for either GAPD or LDH. This may be due to

Table 13. ORD Constants of 6-Phosphofructokinase at
Different pH's and During Its Interaction with ATP

Constants	PFK pH 8.5	PFK pH 6.1—7.1	PFK + ATP, pH 8.5		PFK + ATP, pH 6.5	
			2.5 mM ATP	5 mM ATP	2.5 mM ATP	5 mM ATP
K_c 10^{-7} (±0.03)	—0.55	—0.7	—0.52	—0.5	—0.75	—0.55
λ_c,mμ ($\pm$3)	287	277	287	284	250	252
a_0 ($\pm$10)	—86	—120	—90	—78	—130	—90
b_0 ($\pm$10)	—220	—180	—190	—190	—84	—87
A_{193}	+364	+310	+330	+340	+197	+190
A_{235}	—340	—342	—345	—320	—260	—260

localization of the α-helical segments in the hydrophobic interior
of the molecule.

The influence of pH on the conformations of LDH and GAPD
will be discussed below.

The enzyme 6-phosphofructokinase (PFK) catalyzes the conversion of fructose-6-phosphate (F-6-P) to fructose-1,6-diphos-
phate (F-1,6-P). The phosphorus residue is transferred to the
substrate from adenosine triphosphate (ATP):

$$PFK + F-6-P + ATP \rightleftharpoons PFK + F-1,6-P + ADP,$$

where ADP is adenosine diphosphate. It has been proven that PFK
is an allosteric enzyme (see Chapter 12), whose activity is regulated by adenosine monophosphate (AMP), ADP, ATP, and other
phosphoric acid compounds [189]. ATP is an allosteric inhibitor
(see p. 123) of PFK at pH 6.3-7.1 but does not inhibit this enzyme
at pH 8.0-9.0. There are data indicating the PFK has a complex
quaternary structure that is altered under the influence of ATP.
Study of the ORD has made it possible to detect changes in the
secondary and tertiary structure of the enzyme [189, 190]. The
appropriate results are presented in Table 13. The ORD was
studied in PFK solutions with concentrations of 0.1-0.3%, in the
presence and absence of 2.5-5 mM ATP at pH 6.1- 7.1 and 8.5-9.0.
As is shown by Table 13, the constants λ_c and b_0 for PFK alone
(without ATP) differ at different pH's. This indicates that the PFK

has different structures that are sensitive (at pH 6.5) and insensitive (at pH 8.5) to allosteric inhibition. ATP causes no changes in λ_c and b_0 at pH 8.5, where there is no inhibition. On the other hand, it produces substantial changes in these constants, corresponding to a decrease in α-helicity, at pH 6.5. These changes develop gradually, over a period of about an hour. When PFK is preliminarily inactivated with 5.5 M urea, sodium dodecyl sulfate, or 0.1 M NaOH, addition of ATP does not cause changes of this sort.

All the results described above are direct proof that a substrate, coenzyme, inhibitor, or allosteric inhibitor can actually cause conformational changes in an enzyme, particularly in its α-helicity. Measurement of the optical rotatory dispersion is a direct technique for detecting conformational changes. Processing of ORD data by the method of Blout and Shechter makes it possible to obtain information on both the α-helicity and tertiary structure of a protein. A change in effective dielectric constant indicates unfolding of the globule.

The conformational changes observed in dehydrogenases and phosphofructokinase are consistent with the theories discussed in Chapter 6. It has been demonstrated that effectors actually cause changes in enzyme structure. However, it has not been shown that such changes are due to achievement of structural correspondence between the enzyme and the effector. This is far more difficult to do and the only direct method presently suitable for the purpose is x-ray diffraction analysis (see p. 58).

There is some reason to surmise that a method based on the rotation of the polarization plane caused by a magnetic field (the Faraday effect) rather than natural rotation will be most useful for studying the structure of enzymes and other biopolymers. Research on magnetic rotation has only just begun. Magnetic rotation and its anomalous dispersion will be discussed in Appendix II.

Chapter 10

CONFORMATIONAL PROPERTIES OF ENZYMES AND IONIZATION

An enzyme macromolecule contains ionizable groups — cationic and anionic amino acid residues. Ionizable groups may also be present in coenzymes, substrates, inhibitors, and activators, i.e., in any effector. Enzymatic activity thus depends to a large extent on medium pH and hence on the degree of ionization of the corresponding groups. Inhibition and activation of an enzyme also depend on pH. The aforementioned bell-shaped curve dependence on pH is usually observed for enzymatic activity (see p. 87): activity is maximal at some intermediate pH characteristic of the enzyme in question and decreases at larger or smaller pH's, although it does not reach the values corresponding to alkaline and acidic denaturation of the protein.

Webb gives the following classification for the possible mechanisms by which pH affects an enzyme [191].

1. Direct change in the state of the ionizable groups in the active center. (a) Influence on binding of substrates, activators, and coenzymes; (b) influence on rate of decomposition of enzyme—substrate complex; (c) influence on catalytic reaction, if an H^+ ion removed from or attached to the active center participates in it.

2. Indirect influence of the ionizable groups in the active center. (a) Action of ionizable groups adjoining active center; (b) influence of change in total charge on protein.

3. Change in degree of ionization of nonenzymatic components of system. (a) Components directly participating in reaction

94

(substrates, activators, or coenzymes); (b) modifiers indirectly influencing reaction rate (buffers and possible contaminants).

4. Changes in protein structure. (a) Local changes in structure that alter the configuration of the active center; (b) denaturation processes involving the overall structure of the protein; (c) subunit association—dissociation reactions (changes in quaternary structure).

Many of these mechanisms are difficult to distinguish from one another. Webb's book describes methods based on study of the steady-state kinetics of enzymatic reactions. These generally provide only indirect information regarding the influence of pH on an enzyme. It is necessary to conduct investigations by direct physical techniques similar to those described above.

We will consider principally the fourth group of mechanisms in the list given above. This enumeration is essentially incomplete, since it does not contain what may be the most important mechanism of all, i.e., the influence of pH on the higher structural levels of a protein, which is not accompanied by true denaturation. However, we will first consider the general theory of ionization of enzymatic systems and processes belonging to the first and second groups. We will not go into greater detail about the mechanism, limiting ourselves to general considerations.

An enzyme is a polyelectrolyte and can therefore exist in the most diverse ionization states, which are in labile equilibrium with one another. The number of forms with different ionization states obviously depends on pH, but the question is the manner of the dependence. Let us consider the very simple model consisting of a system that can have two ionization states and can accordingly exist in three forms, one unionized and two ionized [193]. A dibasic acid AH_2, such as succinic acid, can serve as an example of such a system:

$$AH_2 \rightleftarrows AH^- \rightleftarrows A^{--}$$
$$K_1 \qquad K_2$$

Here K_1 and K_2 are the first and second ionization constants. The sum of the concentrations of all three forms is

$$A_t = [AH_2] + [AH^-] + [A^{--}]. \qquad (67)$$

The constants K_1 and K_2 are defined by the expressions

$$K_1 = \frac{[AH^-][H^+]}{[AH_2]}, \quad K_2 = \frac{[A^{--}][H^+]}{[AH^-]} = \frac{[A^{--}][H^+]^2}{[AH_2]K_1}. \tag{68}$$

Consequently,

$$A_t = [AH_2]f = [AH^-]f^- = [A^{--}]f^{--}, \tag{69}$$

where f, f^-, and f^{--} are the so-called Michaelis pH functions:

$$\left. \begin{array}{l} f = 1 + \dfrac{K_1}{[H^+]} + \dfrac{K_1 K_2}{[H^+]^2}, \\[2mm] f^- = 1 + \dfrac{[H^+]}{K_1} + \dfrac{K_2}{[H^+]}, \\[2mm] f^{--} = 1 + \dfrac{[H^+]}{K_2} + \dfrac{[H^+]^2}{K_1 K_2}. \end{array} \right\} \tag{70}$$

Let us assume that the active center of an enzyme is reactive only in the intermediate state AH^-. The reaction rate is then

$$v = k[AH^-], \tag{71}$$

where k is a constant, and we obtain

$$v = \frac{kA_t}{1 + \dfrac{[H^+]}{K_1} + \dfrac{K_2}{[H^+]}}. \tag{72}$$

The dependence of v on $[H^+]$, i.e., on pH, passes through a maximum satisfying the condition

$$[H^+] = \sqrt{K_1 K_2},$$

or

$$pH = \frac{1}{2}(pK_1 + pK_2). \tag{73}$$

The theory thus gives a general phenomenological explanation of the processes observed and leads to formulas for calculating and interpreting the experimental data [191, 192]. However, it is inadequate to establish the true molecular nature of these phenomena.

Kirkwood and Shumaker gave a somewhat different interpretation of the dependence of reaction rate on pH [60, 194]. They considered the fluctuations in the electric charges in the enzyme molecule. Since an enzyme contains many ionizable groups and the free energies of the different ionization states differ little, charges can be shifted in the enzyme molecule and its active center, and hence in the enzyme—substrate complex. Charge fluctuations can lead to appearance of an additional electrostatic interaction between the enzyme and substrate, i.e., to development of bonds between them. This theory thus sets out not to explain the dependence of enzymatic activity on pH but to give a general interpretation of such activity: the gain in fluctuational energy during the enzyme—substrate interaction should lead to a decrease in the effective activation energy of the process (see p. 43).

Let U be the potential energy of the enzyme—substrate system at a given charge distribution. This energy should be averaged over all distributions in accordance with the formula

$$e^{-W/kT} = \langle e^{-U/kT} \rangle; \tag{74}$$

where $\langle \ \rangle$ is the average sign, and W is the averaged energy. We obtain

$$W = \langle U \rangle - \frac{1}{2kT} [\langle U^2 \rangle - \langle U \rangle^2] + \cdots \tag{75}$$

In the absence of fluctuations, $\langle U^2 \rangle = \langle U \rangle^2$ and $W = \langle U \rangle$. Let the protein contain n basic groups carrying the charges $z_j e$; designating the dipole moment of the substrate as p, the distance between the i-th group of the active center and the substrate as R_i, and the dielectric constant of the medium as ε, we find

$$U = \sum_{i=1}^{n} \frac{(z_i + x_i) \, ep \cos \gamma_i}{\varepsilon R_i^2}, \tag{76}$$

where γ_j is the angle between R_i and p. The value of x_i is one or zero, depending on whether or not a proton is attached to the i-th group. Let us calculate W.

We have

$$\langle U \rangle = \sum_{i=1}^{n} \frac{z_i + \bar{x}_i}{\varepsilon R_i^2} \, ep \cos \gamma_i, \tag{77}$$

$$\bar{x}_i = 0 \frac{[A_i^-]}{[A_iH] + [A_i^-]} + 1 \frac{[A_iH]}{[A_iH] + [A_i^-]} = \frac{[A_iH]}{[A_iH] + [A_i^-]} = \frac{[H^+]}{[H^+] + K_i}, \quad (78)$$

where

$$K_i = \frac{[A_i^-][H^+]}{[A_iH]}. \quad (79)$$

The mean square energy is written in the form

$$\langle U^2 \rangle = \sum_i (z_i^2 + 2z_i\bar{x}_i + \bar{x}_i^2) \frac{e^2 p^2 \cos^2 \gamma_i}{\varepsilon^2 R_i^4}$$
$$+ \sum_{i,j}' \overline{(z_i + x_i)(z_j + x_j)} \frac{e^2 p^2 \cos \gamma_i \cos \gamma_j}{\varepsilon^2 R_i^2 R_j^2}, \quad (80)$$

while

$$\overline{x_i^2} = \bar{x}_i^2, \quad \overline{x_i x_j} = \bar{x}_i \cdot \bar{x}_j = \frac{[H^+]^2}{([H^+] + K_i)^2}, \quad (81)$$

since groups i and j are identical. We obtain

$$\langle U^2 \rangle - \langle U \rangle^2 = \sum_i (\overline{x_i^2} - \bar{x}_i^2) \frac{e^2 p^2 \cos^2 \gamma_i}{\varepsilon^2 R_i^4}, \quad (82)$$

where

$$\overline{x_i^2} - \bar{x}_i^2 = \frac{[H^+]}{[H^+] + K_i} \left(1 - \frac{[H^+]}{[H^+] + K_i} \right) = \frac{[H^+] K_i}{([H^+] + K_i)^2}. \quad (83)$$

Thus,

$$W = \sum_{i=1}^{n} \left[z_i + \frac{[H^+]}{[H^+] + K_i} \right] \frac{ep \cos \gamma_i}{\varepsilon R_i^2}$$

$$- \frac{1}{2kT} \sum_{i=1}^{n} \frac{K_i [H^+]}{([H^+] + K_i)^2} \frac{e^2 p^2 \cos^2 \gamma_i}{\varepsilon^2 R_i^4}. \quad (84)$$

Let the geometry of the system be such that the first sum reverts to zero. We then have only the energy associated with charge fluctuations, i.e.,

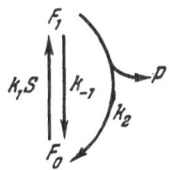

Fig. 27. Scheme of
simple enzymatic re-
action.

$$W = -\frac{n_\alpha e^2 p^2}{4\epsilon^2 r_\alpha^4 kT}\frac{K_\alpha [H^+]}{([H^+]+K_\alpha)^2}, \qquad (85)$$

where

$$\frac{1}{2r_\alpha^4} = \frac{1}{n_\alpha}\sum_{i=1}^{n_\alpha}\frac{\cos^2\gamma_i}{R_i^4}.$$

Here, n_α is the number of adjacent basic groups. In order to make the calculation, we must know the effective dielectric constant, i.e., its local value in the enzyme—substrate contact region. This value is naturally far less than $\epsilon = 80$, the dielectric constant of water. According to Kirkwood's estimate, $\epsilon \approx 10$.

In the simplest case, the steady-state kinetics of an enzymatic reaction, are described by the Michaelis—Menten equation†:

$$v = \frac{k_2ES}{K+S}, \qquad (86)$$

where E is the enzyme concentration and S is the substrate concentration. The sense of the constants k_2 and K follows from the derivation of this equation. Figure 27 is a diagram of the process. F_0 is the free enzyme and F_1 is the enzyme attached to a substrate molecule. Under steady-state conditions,

$$F_1 = k_1SF_0 - (k_{-1} + k_2)F_1 = 0 \qquad (87)$$

and

$$E = F_0 + F_1. \qquad (88)$$

We obtain Equation (86) with $K = (k_{-1} + k_2)/k_1$. K is the equilibrium constant for formation of the enzyme—substrate complex and depends exponentially on the effective energy, i.e.,

$$\ln K = -\frac{W_{\text{eff}}}{kT}. \qquad (89)$$

---

†Henceforth, the reagents and their concentrations are given the same symbols for simplicity.

The charge fluctuations cause the energy and K to vary. The same is true of the rate constant of the reaction k_2; actually,

$$\ln \frac{k_2}{k_2^0} = - \frac{\Delta W^* - \Delta W_0^*}{kT},$$ (90)

where ΔW^* and ΔW_0^* are the activation energies of the reaction in the presence (k_2) and absence (k_2^0) of charge fluctuations. In accordance with the previous calculation,

$$\ln \frac{k_2}{k_2^0} = \frac{n_\alpha e^2 \Delta p^2}{4\varepsilon^2 k^2 T^2 r_\alpha^4} \frac{K_\alpha [H^+]}{([H^+] + K_\alpha)^2},$$ (91)

where Δp^2 is the difference in the squares of the dipole moment of the substrate in the activated, transition state and in the inactivated state.

Equation (91) gives a bell-shaped curve for k_2 as a function of pH. The maximum in k_2 corresponds to $[H^+] = K_\alpha$, i.e., pH = pK_α.

Kirkwood assumed the mechanism of the fluctuation interaction to be universal and attributed it both to mutual attraction of the enzyme and substrate and to interaction of the protein molecules themselves during their aggregation [195]. If two protein molecules are a certain distance apart, the electric field of the fluctuating charge distribution of one molecule alters the charge distribution in the other molecule in such fashion that attractive forces appear. These forces are nonspecific. Timasheff recently devised a theory of protein interaction caused by proton fluctuations [196]. He demonstrated that a substantial attractive force caused by such fluctuations can develop between identically ionized groups at pH's close to their pK_α. The attractive energy reaches 1-2 kcal/mole in the dielectric medium at the surface of a protein molecule ($\varepsilon \sim 20$). The dependence of the attractive energy on pH has the form of a bell-shaped curve. Timasheff compared the results of his calculations with data on the specific aggregation of α-chymotrypsin and β-lactoglobulin and obtained satisfactory agreement between the theoretical and experimental results.

The calculations described above take no account at all of the cooperativity of charge interactions. Charge fluctuations must evidently exist and, as Timasheff correctly noted, they cannot be ignored. On the other hand, there is no reason to assume that the bell-shaped curve dependence of enzymatic activity or aggregation energy on pH is due to charge fluctuations. Other factors, including the multibasicity of a polyelectrolyte (in conformity with the theory advanced by Michaelis and Davidsohn) and conformational phenomena (see below) can play an equal, if not more important, role. No direct proof of the existence of charge fluctuations has yet been obtained. Kirkwood and Shumaker thought that such fluctuations should lead to appearance of additional components in the relaxation-time spectrum. Sheider [197] recently demonstrated that this is not so. It is obvious that the theories devised by Michaelis and Kirkwood are not related to current concepts of the conformational lability of enzymes. Moreover, induced enzyme—substrate contact is effected by bond rotations, which are accompanied by charge displacement. Electrostatic interactions (including formation of ionic pairs and salts) between the side chains can play as great a role as the state of the hydrogen bonds in α-helices. Displacement of the hydration spheres surrounding the ionized groups is also important. In any event, the importance of establishing the extent to which the influence of pH on enzymatic activity is associated with its influence on enzyme conformation is obvious.

There is at present no theory to account for the influence of pH on the three-dimensional structure of proteins. Study of the hydrophobic interactions that shape the globule (see Chapter 3) as a function of medium pH must be regarded as very urgent. No work has yet been done on this problem.

However, it is possible to make a rigorous analysis of the influence of medium pH on the helicity of a protein, i.e., on its secondary structure. Appropriate model calculations have been made in articles by Vol'kenshtein and Fishman [198, 199]. If a helical polyamino acid consists of ionizable groups, the helix—random coil transition occurs at pH's close to the pK's of these groups. The appearance of charges on the units of the chain causes mutual repulsion and the helix unfolds. Appropriate statistical thermodynamic calculations for polyglutamic acid were given by Zimm and Rice [200] (see also [4]). The statistical sum for such a polyelectrolyte chain has the form

$$Z = \sum_{\{\mu_i\}} \exp\left[-\frac{F(\{\mu_i\})}{kT}\right] \sum_{\{\eta_i\}} \prod_{i=1}^{N} a^{\eta_i} \exp\left[-\frac{F_{\{\mu_i\}}^{(e)}(\{\eta_i\})}{kT}\right]. \qquad (92)$$

Here μ_i equals one when a hydrogen bond is present in the i-th unit stabilizing the chain and zero when such a bond is absent; n_i is one when there is a charge on the i-th unit and zero when there is none. $F\{\mu\}_i$ is the free energy of the uncharged chain with a given set of states $\{\mu_i\}$; $F^{(e)}_{\{\mu_i\}}(\{\eta_i\})$ is the free energy of the electrostatic interaction of the charged groups, which depends on the sets of $\{\mu_i\}$ and $\{\eta_i\}$; summation is carried out over all the states of the chain $\{\mu_i\}$ and $\{\eta_i\}$; N is the number of units in the chain, and a is the ratio of the activities of the charged and uncharged units, expressed in terms of the pH and pK in the following manner:

$$\log a = \pm (\text{pH} - \text{pK}); \qquad (93)$$

the plus and minus signs are for acidic and basic groups, respectively.

The helix—random coil transition is a cooperative process having much in common with the melting of a crystal [3, 4]. Rupture of a single hydrogen bond does not lead to a gain in free energy at the transition point, since enthalpy is expended and the entropy remains unchanged; a single unit of the helix cannot be liberated. Conversely, rupture of several consecutive hydrogen bonds leads to liberation of residues. The process is therefore cooperative. In Equation (92), the cooperativity is expressed by the specific dependence of $F(\{\mu_i\})$ and $F^{(e)}_{\{\mu_i\}}(\{\eta_i\})$ on the values of μ_i and η_i. The state of each unit depends on that of the adjacent units. Each unit correspondingly introduces into the partition function the factor 1 if $\mu_i = 0$, the factor

$$s = e^{-\Delta F/kT},$$

if $\mu_i = 1$, the factor

$$\sigma = \exp\left[-\frac{F_{\text{init}}}{kT}\right]$$

for each monomeric unit in the state $\mu_i = 1$ following three or more unbonded units $(\mu_{i-1} = \mu_{i-2} = \mu_{i-3} = 0)$, and the factor 0 for each unit with $\mu_i = 1$ following fewer than three unbonded units; s is the equilibrium constant for the hydrogen-bond formation reaction in a monomer following a bonded monomer $(\mu_{i-1} = 1)$, while σ is the cooperativity parameter (the equilibrium constant for rupture of one bond in a sequence of hydrogen bonds). There is no cooperativity if $\sigma = 1$ and cooperativity is maximal if $\sigma = 0$, i.e.,

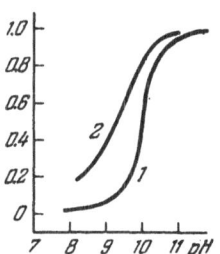

Fig. 28. Curves representing helicity (1) and titration, i.e., ionization (2), of polylysine in water.

the helix as a whole should immediately be converted to a random coil. The foregoing pertains to a nonpolyelectrolyte chain, i.e., the first sum in Equation (92). In the case with which we are concerned, the situation is complicated, but the partition function can nevertheless be computed. Calculations have shown that the transition point characterized by the value of s is displaced under the action of electrostatic forces and thus of a change in pH. In a polyamino acid, which contains only one type of residue, the helix—random coil transition occurs within a narrow pH range. The curve representing helicity as a function of pH runs roughly parallel to that representing the molecular ionization (titration curve). This is shown in Fig. 28 [201].

If a polypeptide chain contains both acidic and basic groups, dehelicization can occur at both acidic and alkaline pH's.

In some cases, the helix proves to be stable at neutral pH's, since the ionizable groups of both types are charged in this region and the electrostatic energy of their attraction reduces the free energy of the system.

Let us consider the simplest model, which consists of a copolymer with one ionizable residue for every three nonionizable amino acid residues; the acidic and basic residues then alternate regularly (Fig. 19) [198]. Such a copolymer is obviously a better model of a protein than a homogeneous polyamino acid. Since the charges are rather far apart along the chain, we can limit ourselves to the nearest-neighbor interaction in zero-order approximation. Calculation of a partition function of the type of Equation (92) for this model actually yields a bell-shaped curve for the dependence of helicity on pH. A rectangular curve (Fig. 30) is obtained for complete cooperativity ($\sigma = 0$); the axis of symmetry intercepts the abscissa at the point pH $= \frac{1}{2}(pK_A + pK_B)$. The width of the rectangle or bell depends to a large extent on s and thus on the temperature. Investigation of the temperature function of bell width can thus provide information on the value of ΔF, i.e., the free energy

Fig. 29. Model of ionizable copolymer chain.

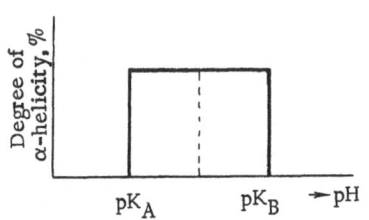

Fig. 30. Degree of α-helicity as a function of pH for model polymer.

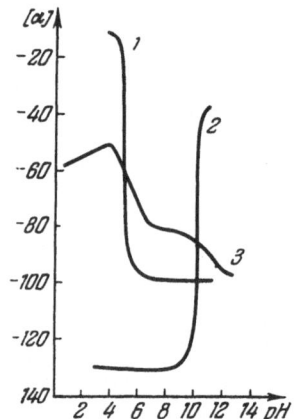

Fig. 31. Specific rotation as a function of medium pH. 1) Poly-l-glutamic acid; 2) poly-l-lysine; 3) copolymer.

of hydrogen-bond formation in a monomer following a bonded monomer. No appropriate experimental investigations have as yet been conducted.

Generally speaking, the curve representing the helicity of a protein as a function of pH can also fail to be bell-shaped. This function is governed by the mutual positioning and relative content of acidic and basic groups in the protein. Vol'kenshtein and Fishman considered different models of the polypeptide chain [198]. They showed that, if the chain contains sufficiently large units consisting of the same amino acid, the curve may be concave rather than convex. It is also possible for two maxima located near the pK's of the corresponding groups to appear in the curve. Detailed study of the dependence of helicity on pH can thus provide valuable information on protein structure. The problem is naturally complicated by the existence of electrostatic interactions not directed along the chain in the tertiary structure of the protein.

The theory developed in the aforementioned article [198] accounts for the dependence of the helicity of muscle proteins on pH, which was studied by Lowey [202]. It proved possible to establish a correlation between the trend of the curve and the relative content of anionic and cationic residues in tropomyosin and other muscle proteins.

Doty et al. [203] and Blout and Idelson [204] experimentally determined the helicity of copolymers of l-lysine and l-glutamic acid, at different pH's. Figure 31 shows the specific rotations of polylysine, polyglutamic acid, and their copolymer as a function of pH. Vol'kenshtein and Fishman were able to make calculations that agreed with these experimental data [199]. They computed the electrostatic energy of the copolymer, demonstrating that its dependence on pH results from the difference in the electrostatic interaction of the charges on the monomers, which is produced by the difference in side-chain length. These calculations made it possible to obtain reasonable values for the length of the cooperative chain segment and the change in free energy in the uncharged chain during transition of a unit from the randomly coiled to the helical state. A theory of the influence of pH on the α-helicity of proteins that was open to experimental verification was thus created. One direct research technique that can be employed here is measurement of the ORD at different pH's and establishment of the correlation between the dependence of helicity on pH and that of enzymatic activity on this factor. We previously mentioned the results of a study of α-chymotrypsin (see p. 87). More conclusive data were obtained in studying dehydrogenases.

Investigation of the ORD of lactic dehydrogenase (LDH, see p. 88) at different pH's [179] indicated that the dependence on pH of the constants λ_c and K_c in the Drude and Moffitt equations (Fig. 32) follows a bell-shaped curve. The maximum values of λ_c and b_0 occur at neutral pH's (6.0–8.7); they decrease at larger or smaller pH's. The change in the ORD constants is reversible at pH's between 4.0 and 10.6.

Figure 32 shows curves representing the rate of the forward and reverse reactions (pyruvate $\rightleftharpoons$ lactate) as a function of pH. Comparison of these curves with the ORD curve shows that there is a correlation between enzymatic activity and enzyme α-helicity for the pyruvate—reduction reaction. The maximum rate corresponds to the same pH values as the maximum λ_c. This correlation is substantially better for LDH in the presence of the coenzyme NAD, which alters the secondary structure of LDH (see p. 88), the changes depending on pH (see p. 27). The reversible con-

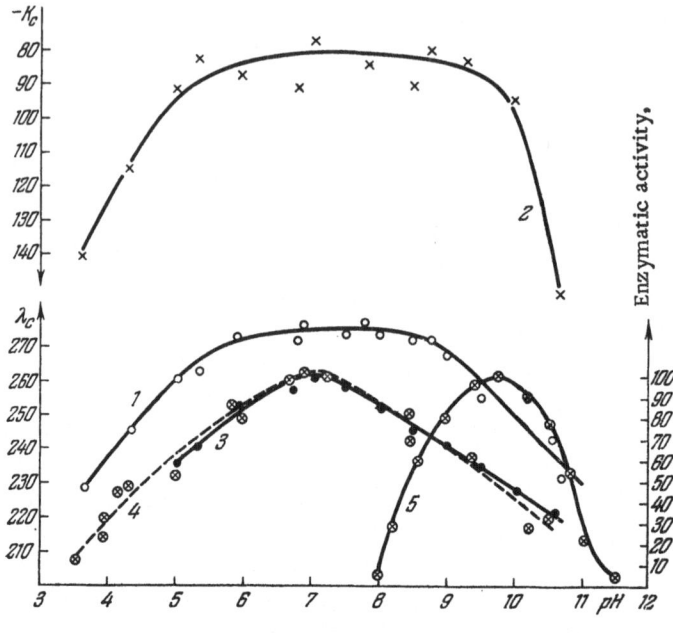

Fig. 32. Constants in Drude equation and enzymatic activity of LDH as functions of pH. 1) λ_c as a function of pH for LDH in 0.06 M phosphate buffer; 2) K_c as a function of pH for LDH; 3) λ_c as a function of pH for LDH + NADH; 4) enzymatic activity of LDH as a function of pH for reduction of sodium pyruvate to lactate ($3.3 \cdot 10^{-4}$ M pyruvate + $1.77 \cdot 10^{-4}$ M NADH + 0.006 M phosphate buffer + $2.2 \cdot 10^{-9}$ M LDH); 5) the same, for reverse reaction ($2.5 \cdot 10^{-5}$ M d,l-lactate + $2.2 \cdot 10^{-2}$ M LDH + $5.4 \cdot 10^{-7}$ M NAD + 0.1 M glycine buffer).

formational changes in LDH when the pH is varied are accompanied by reversible changes in reaction rate.

Analysis of the ORD by the method of Blout and Shechter shows that the experimental points fall on a line representing the dependence of A_{225} on A_{193}, which corresponds to $\varepsilon < 30$, i.e., an organic solvent. This indicates that the α-helical segments are in a hydrophobic environment. The experimental points are displaced along this line when the pH is varied. Hence, it can be concluded that the changes in α-helicity caused by changes in pH are not accompanied by unfolding of the globule, i.e., a fundamental change in the tertiary structure of LDH.

There is no correlation with conformational transformations for the lactate—oxidation reaction.

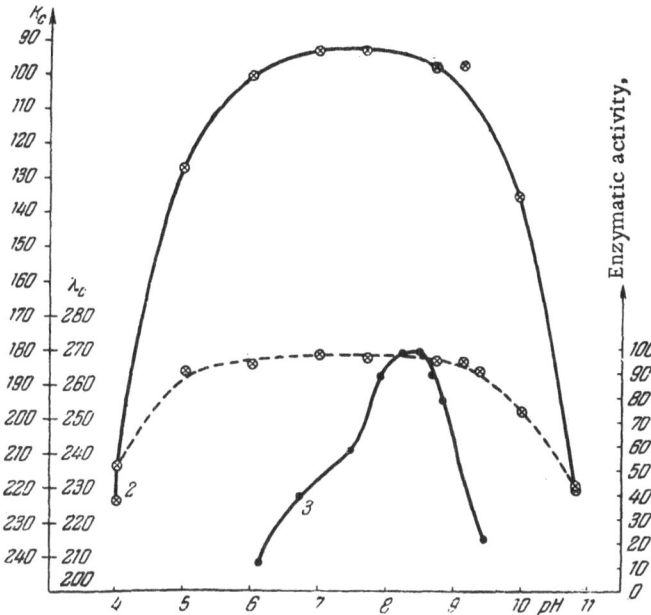

Fig. 33. Constants in Drude equation and enzymatic activity of GAPD as functions of pH. 1) λ_c as a function of pH for GAPD with 2.5 M NAD in 0.1 M glycine buffer; 2) K_c as a function of pH; 3) enzymatic activity as a function of pH during oxidation of d-glyceric aldehyde (7 mM substrate, 1.66 mM NAD, 0.007 μM GAPD, and 0.1 M glycine buffer).

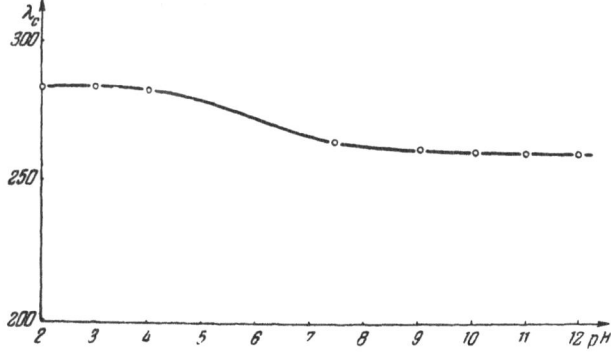

Fig. 34. Dependence of λ_c on pH for insulin.

Investigation of the ORD of d-glyceraldehyde-3-phosphate de-hydrogenase (GAPD) yielded similar results [183] (Fig. 33). Over the pH range 6.0-9.0, GAPD has a stable secondary structure char-acterized by rather high values of λ_c and b_0. The tertiary struc-ture, which is to some extent governed by the constant K_c, is evi-dently more labile: changes in K_c are observed at a constant λ_c.

Figure 33 also gives a curve for the rate of the reaction de-scribed on p. 88 as a function of pH. In contrast to LDH, there is no complete correlation between the enzymatic activity of the GAPD and its α-helicity. The experimental points lie on the Blout—Shechter line corresponding to $\varepsilon < 30$ at pH's between 6 and 9; the points are displaced toward the line corresponding to higher pH's when the pH is varied and, at extreme pH's (4.0 and 10.9), reach the line corresponding to an aqueous solution ($\varepsilon > 30$). Unfolding of the globule consequently occurs at extreme pH's. The tertiary structure of GAPD is more labile than that of LDH.

The results presented above confirm the general theories of the decisive role of conformational changes in enzymatic activity. The correlation found for LDH is very indicative in this sense. There are no grounds for excluding the possibility that pH affects enzymatic activity governed by other factors. It must therefore be assumed that so complete a correlation as that found for LDH is the exception rather than the rule. The lack of a complete correla-tion for GAPD should be no surprise.

The above results are also in agreement with the theory con-sidered on p. 103. It would be interesting to analyze them on the basis of the amino acid compositions of LDH and GAPD. The re-sults obtained by Bolotina [183] are confirmed by those of Listovsky et al. [184].

In concluding this chapter, we will discuss the results ob-tained in Bolotina's study of insulin [206]. Both the amino acid composition and residue sequence, i.e., the primary structure, are known for this comparatively simple protein ([3], p. 116). Cal-culations based on the theory in question fail to yield a bell-shaped curve for the α-helicity of insulin as a function of pH. The experi-mental results shown in Fig. 34 are in agreement with the calcula-tions.

Other data on the influence of medium pH on the secondary structure of different proteins are given in Joly's monograph [205] Study of the ORD of proteins has thus proved that the medium pH has a direct influence on their conformation.

Chapter 11

COOPERATIVE PROPERTIES OF
ENZYMES AND REACTION KINETICS

Enzymes are complex cooperative systems by virtue of their macromolecular structure. All conformational transformations in protein molecules are cooperative in character, so that the development of induced structural correspondence between an enzyme and substrate is itself a cooperative process. In this sense, enzymatic activity is based on the same phenomena that produce the elasticity of rubber, i.e., those of cooperative rotational isomerization [2, 206]. An understanding of this important hypothesis is the starting point for enzyme physics, whose development has only just begun.

As we have seen, we cannot exclude the possibility that, in addition to their geometric and energetic role, conformational transformations are events leading to accumulation of energy in the active center and thus in the substrate.

In accepting the theory of induced structural correspondence, we must emphasize the cooperative character of the enzyme—substrate interaction it organizes. All the bonds participating in this interaction (electrostatic, hydrophobic, hydrogen, etc.) are cooperative in the sense that the state of a given functional group of the protein depends on the state of the neighboring groups, i.e., on their conformation. The active center of an enzyme is a cooperative system.

The macromolecular nature of biologically functional compounds, principally proteins and nucleic acids, is responsible for the important role of cooperative phenomena in molecular-biological processes [3, 207].

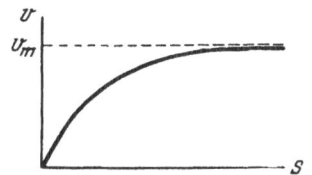

Fig. 35. Curve for v(S) described by Michaelis—Menten equation.

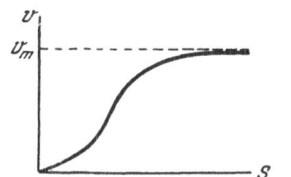

Fig. 36. Curve for v(S) with inflection.

Fig. 37. Diagram of enzymatic process in presence of two equivalent and interacting active centers.

Some enzymes consist of subunits, i.e., have a quaternary structure. Such molecules therefore contain several active centers. It is precisely because protein subunits are macromolecular structures that they can enter into interactions, which also have a cooperative character. Here we are speaking of cooperativity at the level of quaternary structure, which is manifested in the specific kinetics of reactions involving such multicenter enzymes. It was long ago noted that it is the exception rather than the rule for the course of a steady-state enzymatic reaction to obey the simple Michaelis—Menten equation (see p. 99)

$$v = \frac{v_m S}{K + S},\qquad (86a)$$

where v is the reaction rate, v_m is the maximum rate at a substrate concentration S → ∞, and K is a constant. The curve for v(S) described by Equation (86), like the Langmuir isotherm, tends toward saturation without an inflection (Fig. 35). However, the experimental curve for v(S) has an inflection point, i.e., is S-shaped in many cases (Fig. 36), and a maximum is sometimes observed. Attempts have been made to explain these anomalies on the basis of formation of a complex between two or more enzyme molecules and a single substrate molecule [208]. This hypothesis does not stand up to criticism; such a complex could develop only when there was a large excess of enzyme and it is almost impossible to imagine organization of structural correspondence between several enzyme molecules and one substrate molecule. The untenable hypothesis of the existence of such a complex cannot be preserved by a premise of a somewhat differ-

ent type, i.e., that a complex is formed by a single enzyme molecule having several active-center segments, and that the interaction of one segment already bonded to the substrate with another
segment is bimolecular, which also causes the kinetics to deviate
from the Michaelis—Menten equation [209]. This notion is erroneous, since it deals with active segments within a single enzyme
molecule. Their concentration cannot be identical to the enzyme
concentration in the solution and the corresponding reaction is
monomolecular rather than bimolecular. A rigorous calculation
again yields a formula of the type of Equation (86) [210].

The true causes of the appearance of inflections and maxima
in the v(S) curves lie in the cooperative interaction of the protein
subunits, which we define as the interaction of two or more active
centers. Let us consider the simplest case, an enzyme containing
two active centers interacting with one another [210]. Figure 37 is
a diagram of the steady-state process. It is obvious that the enzyme molecule can exist in three states: F_{00}, with both active centers free of the substrate S; $F_{10} = F_{01}$, with one center free and the
other occupied, and F_{11}, with both centers occupied by substrate
molecules. We have the following equations for the steady-state
kinetics:

$$
\left.
\begin{aligned}
\dot{F}_{00} &= -2k_1 S F_{00} + 2(k_{-1} + k_2) F_{10} = 0, \\
\dot{F}_{10} &= k_1 S F_{00} - (k_{-1} + k_2 + k_3 S) F_{10} + \\
&\quad + (k_{-3} + k_4) F_{11} = 0, \\
\dot{F}_{11} &= 2k_3 S F_{10} - 2(k_{-3} + k_4) F_{11} = 0.
\end{aligned}
\right\}
\tag{94}
$$

The total enzyme concentration is

$$
E = F_{00} + 2F_{10} + F_{11}.
\tag{95}
$$

The rate at which the substrate is converted to product is written
in the form

$$
v = 2k_2 F_{10} + 2k_4 F_{11}.
\tag{96}
$$

Solving Equations (94) and (95), we find

$$
v = 2ES \frac{k_2 K' + k_4 S}{K K' + 2K'S + S^2},
\tag{97}
$$

where

$$K = \frac{k_{-1} + k_2}{k_1}, \quad K' = \frac{k_{-3} + k_4}{k_3},$$

or, using the notation $k_4/k_2 = \alpha$ and $K'/K = \beta$,

$$v = 2k_2 ES \frac{\beta K + \alpha S}{\beta K^2 + 2\beta KS + S^2}. \tag{97a}$$

The cooperativity is expressed by the difference of α and β from unity, i.e., the difference in the rate constants for states of the system in which the substrate occupies one or two active centers. Actually, at $\alpha = \beta = 1$ we obtain

$$v = 2k_2 ES \frac{K + S}{(K + S)^2} = \frac{v_m S}{K + S}, \tag{86b}$$

i.e., Equation (86a).

When the condition $\alpha = \beta = 1$ is not satisfied, Equation (97a) indicates that an inflection point, a maximum, or both can appear in the curve for $v(S)$. Mathematical analysis has shown [210] that inflection points, i.e., an S-shaped curve, are possible

at $\alpha < 0.5$ and   any β,

at $1 > \alpha > 0.5$ and $\dfrac{\alpha}{2\alpha - 1} > \beta > \dfrac{\alpha^2}{2\alpha - 1}$,

at $\alpha > 1$ and $\dfrac{\alpha^2}{2\alpha - 1} > \beta > \dfrac{\alpha}{2\alpha - 1}$.

Maxima are possible only at $\alpha < 0.5$ and any β; there are no maxima at $\alpha \geq 0.5$.*

There are regions of values of α and β in which the curve for $v(S)$ has no peculiarities. It can be stated that occurrence of peculiarities in the curve for $v(S)$ always indicates the existence of two or more interacting active centers, but the converse statement does not hold: lack of peculiarities does not mean that there are no interacting centers. In a number of cases, study of the steady-state kinetics of enzymatic reactions thus makes it possible to detect cooperativity at the quaternary-structure level.

---

*My previous article [210] contains an error that I will correct here. The second term in Equation (16) should read $2\alpha KS/(2\alpha - 1)$.

The equilibrium function for saturation of the enzyme with substrate in the two-center system under consideration is expressed by a formula similar to Equation (97), i.e.,

$$\overline{Y} = \frac{2F_{10} + 2F_{11}}{2\,(F_{00} + 2F_{10} + F_{11})} = \frac{K'S + S^2}{KK'S + 2K'S + S^2} = \frac{\beta KS + S^2}{\beta K^2 + 2\beta KS + S^2}. \qquad (98)$$

The saturation function contains only two constants, β and K. At $\beta \neq 1$, the graph of the function $\overline{Y}(S)$ can have an inflection but never a maximum. If $\beta = 1$, i.e., there is no cooperativity,

$$\overline{Y} = \frac{S}{K+S}, \qquad (98a)$$

and the saturation function is similar to the Langmuir isotherm.

The kinetic problem involving two active centers discussed above is easily solved. However, steady-state kinetic problems for enzymes with multiple active centers in the presence of substrates, inhibitors, and other effectors can be very complex. Problems of nonsteady-state kinetics are even more complicated [191, 192, 211, 212].

Appendix I describes a new mathematical method for solution of complex problems of enzyme kinetics, which is based on the so-called theory of graphs. It materially simplifies the solution process.

It is natural that the larger the number of constants k_i governing the kinetics of an enzymatic reaction, the more difficult it is to determine them individually. In this connection, some authors speak of the inexpedience of solving complex problems. This is incorrect. Even without having specific values for the constants, one can draw far-reaching conclusions regarding the mechanism of the process from the general character of the dependence of the reaction rate on the concentrations of the substrate and other effectors. A simple example is Equation (97) and its analysis. Without knowing the specific constants, we are able to establish that the process is cooperative and even to find the number of interacting active centers by investigating the form of the curve for $v(S)$.

Information on the cooperative properties of a protein, which are determined principally by its secondary structure, can theoretically be obtained by investigating the kinetics of proteolysis.

Here the protein figures not as an enzyme but as a substrate decomposed by a proteolytic enzyme.

It is customary to distinguish exopeptidases, which are proteinases that catalyze cleavage of the terminal bond of a polypeptide, from endopeptidases, which are proteinases that act on given internal bonds in the chain. The cleavage process can be single-chain or multi-chain. In the first case, the molecule of the biopolymer being hydrolyzed remains in the enzyme—substrate complex until it has been completely decomposed. In the second case, each bond rupture is independent: the ES complex breaks up after hydrolysis of a bond and then reforms with the previous S chain or a new one. Other kinetic mechanisms are also possible. There are reactions in which intermediate hydrolysis products are formed, as well as "explosive" proteolytic reactions that cause rapid degradation of the chain to monomers without formation of noticeable amounts of intermediate oligomers [213]. It is obvious that an "explosive" reaction is more probable with the single-chain mechanism. Exopeptidases and endopeptidases should obviously split peptide bonds in ordered and disordered regions at different rates. Hydrolysis of a helical section is cooperative: the rate at which a given bond is split should depend on the state of the adjacent bonds and proteolysis should therefore proceed simultaneously with denaturation. Is a proteinase not a denaturase [213]? The kinetics of proteolysis can accordingly provide information on the helicity of a protein and even on its tertiary and quaternary structure. However, the theory of these processes has still not been sufficiently well worked out and there are only a few articles in which a start has been made [214, 215]. There have been no sufficiently thorough experimental investigations.

We have seen that detailed study of the steady-state kinetics of enzymatic processes can provide information on the cooperative interactions in the enzyme molecule. This information is naturally phenomenological in character. The nature of the interaction cannot be determined by this method. It would be of assistance to know the rate constants, but it is difficult to count on complete quantitative determination of these factors in any complex situation. In addition to kinetic investigations, studies made by direct and indirect physical and physicochemical techniques are therefore especially important. Molecular biophysics should specifically obtain an answer to the vital question of the mechanism of the interaction of active centers in different subunits of an enzyme molecule. Despite the fact that the distance between such active centers may be great (see Chapter 13), they nevertheless interact. Why is this so? Does it result from conformational rearrangement of each subunit or from some other process?

Especially important data on such phenomena are yielded by research on allosteric enzymes and similar proteins, particularly hemoglobin. The two following chapters are devoted to these proteins.

A change in subunit conformation caused by attachment of a substrate or other effector can obviously produce a change in conformation and thus in the affinity for the substrate in adjacent subunits. One possible mechanism is specific electrostatic interaction. Another conceivable mechanism is a resonance process.

In a number of cases, there are solid grounds for considering the quaternary structure of a "cooperative" protein to be symmetric and consist of identical subunits (see p. 125). All oscillations of such subunits, both acoustic and optical, will be in resonance. The subunits are in contact with one another both directly and through the layer of water molecules between them. Oscillations excited in a given globule by thermal movement thus interact with oscillations in adjacent globules. According to Shnol' [69], it is possible to detect synchronous acoustic oscillations of protein molecules in aqueous solution (see p. 44).

The oscillation interaction produces resonance and frequency splitting. Attachment of a substrate to one globule throws it out of resonance. Attachment of this same substrate to an adjacent globule restores the resonance, but at different frequencies. Let us evaluate the corresponding changes in free energy.

For purposes of simplicity, we will limit ourselves to considering two interacting globules simulated by harmonic oscillators. State 1, equivalent to F_{00} (see p. 112), is characterized by the oscillation frequency ν_1. As a result of the resonance interaction of the globules, this frequency is split into $\nu_1 + \Delta\nu$ and $\nu_1 - \Delta\nu$. The free energy of the system is

$$F_1 = kT \ln\left[\left(1 - \exp\left[-\frac{h(\nu_1 + \Delta\nu)}{kT}\right]\right)\left(1 - \exp\left[-\frac{h(\nu_1 - \Delta\nu)}{kT}\right]\right)\right]. \quad (99)$$

State 2, equivalent to F_{01} or F_{10} (see p. 112), develops as a result of a change in the frequency of one of the oscillators (attachment of a substrate). The frequency of the first oscillator remains ν_1 but the frequency of the second is altered, acquiring the value $\nu_2 = \nu_1 + \delta\nu$. The free energy of the system in this state is

$$F_2 = kT \ln\left[\left(1 - \exp\left[-\frac{h\nu_1}{kT}\right]\right)\left(1 - \exp\left[-\frac{h(\nu_1 + \delta\nu)}{kT}\right]\right)\right]. \qquad (100)$$

In state 3, equivalent to F_{11} (see p. 112), the frequencies of both oscillators are altered: identical substrate molecules are attached to them. The frequencies of the first and second oscillators are the same and equal ν_2; as a result of the resonance interaction, they are split into $\nu_2 + \Delta'\nu$ and $\nu_2 - \Delta'\nu$. The free energy of the system is written in the form

$$F_3 = kT \ln\left[\left(1 - \exp\left[-\frac{h(\nu_2 + \Delta'\nu)}{kT}\right]\right)\left(1 - \exp\left[-\frac{h(\nu_2 - \Delta'\nu)}{kT}\right]\right)\right]. \qquad (101)$$

Attachment of the first substrate molecule to the system increases the affinity for a second substrate molecule if the free energy is more greatly reduced during the $2 \to 3$ transition than during the $1 \to 2$ transition, i.e.,

$$F_2 - F_3 > F_1 - F_2. \qquad (102)$$

Let us clarify the conditions under which this relationship is satisfied. We introduce the following notation:

$$a = \exp\left[-\frac{h\nu_1}{kT}\right], \quad x = \exp\left[-\frac{h\Delta\nu}{kT}\right],$$

$$y = \exp\left[-\frac{h\delta\nu}{kT}\right], \quad z = \exp\left[-\frac{h\Delta'\nu}{kT}\right].$$

All these quantities are positive. We obtain

$$\Delta F = F_1 - F_2 = kT \ln\frac{(1 - ax)\left(1 - \dfrac{a}{x}\right)}{(1 - a)(1 - ay)}, \qquad (103)$$

$$\Delta F > 0, \quad \text{if} \quad a - \left(x + \frac{1}{x}\right) > ay - (1 + y).$$

Since $\Delta\nu > 0$, $x < 1$ and $x + 1/x > 2$. Hence,

$$y(1 - a) > 1 - a;$$

since $\nu_1 > 0$ and $a < 1$, we have y > 1, i.e., attachment of a substrate molecule to one of the oscillators causes a decrease in the free energy at $\exp(-h\delta\nu / kT) > 1$ or $\delta\nu < 0$. Let us also evaluate the difference in the free energies of states 2 and 3; we find

$$\Delta F' = F_2 - F_3 = kT \ln \frac{(1-a)(1-ay)}{(1-ayz)\left(1-\frac{ay}{z}\right)}. \qquad (104)$$

Similar analysis of the inequalities shows that, since $\Delta'\nu > 0$ and z < 1, $\Delta F' > 0$ when y > 1. It follows from $\Delta F > 0$ and $\Delta'F > 0$ that $F_1 - F_3 > 0$, or

$$(1 - ax)\left(1 - \frac{a}{x}\right) > (1 - ayz)\left(1 - \frac{ay}{z}\right)$$

and, since $a > 0$ and y > 1,

$$a + z + \frac{1}{z} > ay + x + \frac{1}{x},$$

whence

$$z + \frac{1}{z} > x + \frac{1}{x},$$

i.e., since 1 > x, we have z > 0 and z < x. Consequently,

$$\Delta'\nu > \Delta\nu.$$

The resonance frequency splitting increases as the frequency decreases.

Inequality (102) is satisfied when y > 1:

$$\Delta F' > \Delta F.$$

The model under consideration is thus in qualitative agreement with the increase in the affinity of a given subunit for a substrate if an adjacent subunit has already bound the same substrate. However, this calculation is of purely illustrative value, since it contains no quantitative estimates. We as yet know nothing about the nature of the hypothetical oscillations discussed or about the possibility of substantial changes in affinity for substrates as a result of the mechanism in question. The feasibility of simulating oscillations in globules with sustained oscillators is also hypothetical.

If we are considering acoustic oscillations, these participate in thermal fluctuations, whose energy is dissipated (see also [319, 320]).

Nevertheless, study of the interactions within an enzyme molecule on the basis of a dynamic rather than a static model seems very attractive.

Chapter 12

ALLOSTERIC ENZYMES

In addition to performing their catalytic function, enzymes also play a specific regulatory role. The living cell and the multicellular'organism as a whole are highly advanced self-regulating systems. The occurrence of regulation implies the existence of communications pathways along which appropriate information is transmitted. Moreover, self-regulation requires feedback; the control apparatus must receive information on the state of the system being regulated and alter its action in accordance with the character of these data. The ability to maintain steady-state conditions (temperature, osmotic pressure, functional-compound concentrations, etc., which are collectively referred to as homeostasis) inherent in living organisms could not exist without special feedback mechanisms.

In regulable devices designed by man, i.e., machines, communications are effected by electrical, magnetic, or mechanical signals. It can be assumed that the principal means of regulation in living organisms is chemical signaling at the molecular level. Hormonal regulation in complex organisms has long been known, but it takes place in the multicellular organism as an integral system. What are the regulatory processes in a single cell, specifically a bacterium?

Jacob and Monod discovered the manner in which the genetic function of DNA is regulated in the operon system [3, 216, 217], and in which DNA replication is regulated in the replicon system [3, 218]. The regulatory enzymes or allosteric enzymes (ASE) that effect feedback in the reaction systems leading to synthesis of a number of metabolites were discovered at the same time.

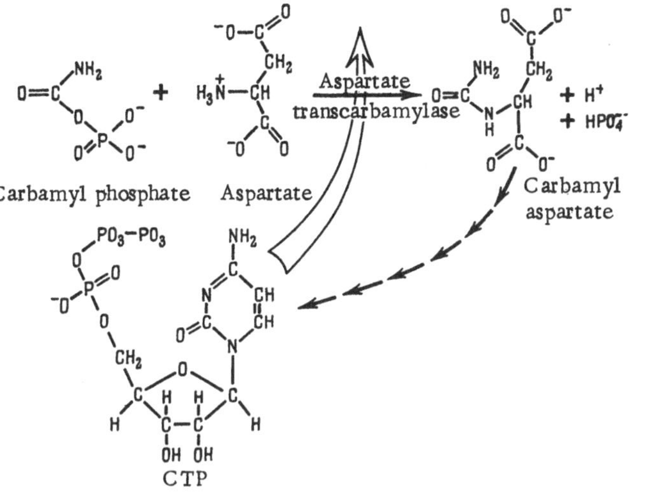

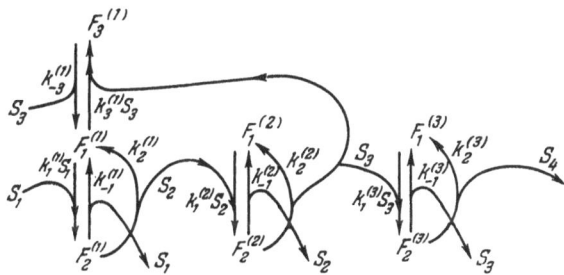

Fig. 38. Diagram of allosteric process involving participation of aspartate transcarbamylase.

Fig. 39. Diagram of allosteric process.

Any metabolite is synthesized in the cell as a result of a
number of successive reactions, each of which is catalyzed by a
specific enzyme. A whole set of enzymes is thus required for
synthesis of a given metabolite. Umbarger was the first to show
that there are reaction sequences in which the final metabolite affects the activity of the enzyme catalyzing the first reaction in the
sequence, despite the fact that this metabolite has nothing in common with the substrate of the enzyme in question [219]. He established that the threonine deaminase of *E. coli*, whose substrate is

threonine, is inhibited by isoleucine, which is the terminal product of the reaction chain. A number of similar processes were subsequently discovered; several dozen ASE are now known (the meaning of the term "allosteric" will be clarified by the following discussion). Figure 38 is a diagram of an allosteric process in which the ASE is aspartate transcarbamylase [220]. Before considering the molecular properties of ASE, let us examine the behavior of the systems they regulate. Figure 39 is a diagram of the process; we can limit ourselves to three successive reactions without loss of generality [221]. The steady-state kinetic equations have the form

$$
\left.
\begin{aligned}
\dot{F}_1^{(1)} &= -(k_1^{(1)}S_1 + k_3^{(1)}S_3)\, F_1^{(1)} + (k_{-1}^{(1)} + k_2^{(1)})\, F_2^{(1)} + k_{-3}^{(1)} F_3^{(1)} = 0, \\
\dot{F}_2^{(1)} &= k_1^{(1)}S_1 F_1^{(1)} - (k_{-1}^{(1)} + k_2^{(1)})\, F_2^{(1)} = 0, \\
\dot{F}_3^{(1)} &= k_3^{(1)}S_3 F_1^{(1)} - k_{-3}^{(1)} F_3^{(1)} = 0, \\
E_1 &= F_1^{(1)} + F_2^{(1)} + F_3^{(1)}, \\
\dot{F}_1^{(2)} &= -k_1^{(2)}S_2 F_1^{(2)} + (k_{-1}^{(2)} + k_2^{(2)})\, F_2^{(2)} = 0, \\
\dot{F}_2^{(2)} &= -\dot{F}_1^{(2)} = 0, \\
E_2 &= F_1^{(2)} + F_2^{(2)}, \\
\dot{F}_1^{(3)} &= -k_1^{(3)}S_3 F_1^{(3)} + (k_{-1}^{(3)} + k_2^{(3)})\, F_2^{(3)} = 0, \\
\dot{F}_2^{(3)} &= -\dot{F}_1^{(3)}, = 0, \\
E_3 &= F_1^{(3)} + F_2^{(3)}.
\end{aligned}
\right\}
\tag{105}
$$

Solving these systems, we find that the rates of the reactions $S_1 \to S_2$, $S_2 \to S_3$, and $S_3 \to S_4$ are, respectively,

$$
\left.
\begin{aligned}
v_1 &= k_2^{(1)}F_2^{(1)} = \frac{v_{m1}S_1}{S_1 + K_1\,(1 + K_I S_3)}, \\
v_2 &= k_2^{(2)}F_2^{(2)} = \frac{v_{m2}S_2}{S_2 + K_2}, \\
v_3 &= k_2^{(3)}F_2^{(3)} = \frac{v_{m3}S_3}{S_3 + K_3},
\end{aligned}
\right\}
\tag{106}
$$

where $v_{mi} = k_2^{(i)} E_i$ (i = 1, 2, or 3) is the maximum rate, $K_i = (k_{-1}^{(i)} + k_2^{(i)})/k_1^{(i)}$ is the Michaelis constant, and $K = k_3^{(1)}/k_{-3}^{(1)}$.

Under steady-state conditions, $v_1 = v_3$ (compare [191]). We obtain

$$S_3^2 + \frac{1}{K_I}\left[1 + \frac{S_1}{K_1}\left(1 - \frac{v_{m1}}{v_{m3}}\right)\right]S_3 - \frac{v_{m1}K_3S_1}{v_{m3}K_1K_I} = 0. \tag{107}$$

It is readily seen that the rate of S_3 synthesis is governed by the consumption of this compound in subsequent reactions, i.e., by the rate at which it is converted to S_4. If v_{m3} is large, S_3 is small and there is no inhibition of the first enzyme. Conversely, if v_{m3} is small, i.e., $v_{m3} \ll v_{m1}$, Equation (107) takes the form

$$S_3^2 - \frac{S_1}{K_1K_I}\frac{v_{m1}}{v_{m3}}S_3 - K_3\frac{S_1}{K_1K_I}\frac{v_{m1}}{v_{m3}} = 0. \tag{108}$$

Since $v_{m1}/v_{m3} \gg 1$, S_3 is large. If $S_3 \gg K_3$,

$$S_3 \approx \frac{S_1}{K_1K_I}\frac{v_{m1}}{v_{m3}} \tag{109}$$

and

$$v_1 \approx \frac{v_{m1}v_{m3}S_1}{(v_{m1} + v_{m3})S_1 + K_1v_{m3}} \ll \frac{v_{m1}S_1}{S_1 + K_1}. \tag{110}$$

The maximum rate of the first reaction is not v_{m1} but v_{m3}, i.e., substantially lower. At $v_{m3} \to 0$, we obtain $S_3 \to \infty$ and $v_1 = v_3 \to 0$, i.e., complete inhibition of the process. It can be seen that the system under consideration is regulated by feedback.

Let us now turn to the molecular properties of ASE. According to the diagram in Fig. 39, the product in the second reaction, which we have designated as S_3, interacts with the enzyme E_1, which is an ASE, as a competitive inhibitor. However, ordinary competitive inhibitors are similar in structure to the enzyme substrate. In the case of an ASE, the allosteric effector (ASEf), the inhibitor S_3, does not bear any such similarity to the substrate S_1. The structural correspondence organized by S_3 consequently differs from that organized by S_1. It can be assumed that S_3 acts on a segment of the ASE molecule that does not coincide with the active center for S_1. At the same time, the ASEf prevents interaction of the ASE with its substrate. This means that the inhibitory action of the ASEf is exerted indirectly rather than directly. Hence, the term "allosteric," which is a literal transcription of an ancient Greek word meaning "spatially different."

Different models have been proposed for the indirect effect. Monod, Changeux, and Jacob [222] initially thought that achievement

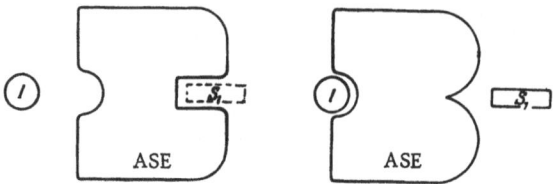

Fig. 40. Model of allosteric inhibition (after Monod, Changeux, and Jacob).

of spatial correspondence with the inhibitor S_3 alters the conformation of the ASE in such fashion that structural correspondence with the substrate S_1 cannot be realized (Fig. 40). Braunshtein described this model still more graphically. Let us imagine an inflated rubber ball, on whose surface is drawn a figure such as a square to simulate the active center for S_1. If we press on some other area of the ball, the square is deformed and the structural correspondence is disrupted.

This model explains important properties of ASE. It is based on the assumption that an ASE molecule as a whole can undergo conformational transformations. We are essentially considering a single cooperative system: a change in the conformation of one part of the system entails conformational transformations in other segments. This model is obviously in agreement with Koshland's hypothesis of induced structural correspondence. If it is valid, allosteric inhibition must be regarded as solid confirmation of this hypothesis. Both Koshland's theory and the model under consideration deal with induced changes in the tertiary and secondary structure of proteins.

The aforementioned cooperativity does not mean that the kinetics of reactions in which ASE participate must be cooperative. The expression we have obtained for v_1 in no way differs from that for noncooperative competitive inhibition.

Experimentation has shown, however, that the kinetics of the corresponding reactions are cooperative. Figure 41 shows the reaction rate as a function of aspartate concentration for aspartate transcarbamylase [220]. The curve has a pronounced S shape. The cooperativity disappears in the presence of an ASEf (the inhibitor CTP). The kinetics of inhibition are in turn cooperative.

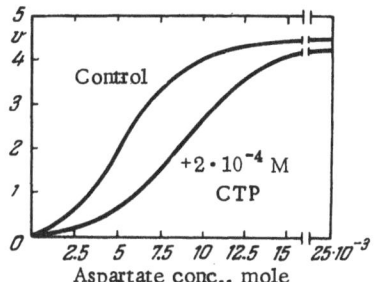

Fig. 41. Function $v(S)$ for aspartate transcarbamylase.

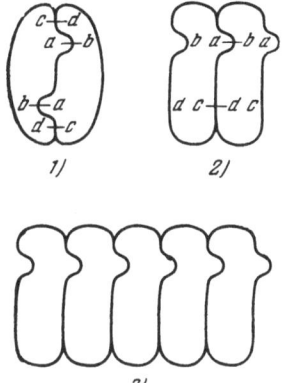

Fig. 42. Diagram of isologous dimer (1),
heterologous dimer (2), and heterologous
polymer (3).

This was especially clearly demonstrated by Changeux, who studied another ASEf, threonine deaminase [223]. The expression for the kinetics of substrate (threonine) conversion has the form

$$\log \frac{v}{v_m - v} = n \log S - \log K \qquad (111)$$

while that for the kinetics of inhibition has the form

$$\log \frac{v}{v_0 - v} = \log K' - n' \log I, \qquad (112)$$

where v is the rate, v_m is the maximum rate, v_0 is the rate at an inhibitor (I) concentration of zero, and K and K' are constants; these expressions are characterized by the coefficients $n > 1$ and $n' > 1$. Changeux obtained $n = 1.37$ and $n' = 1.86$. If the reaction kinetics were noncooperative, n and n' would be one. As we have seen, cooperativity of the type under consideration implies the existence of two or more interacting active centers. In this case, there should be at least two centers for the substrate and two for the ASEf. Actually, calculations based on this model for a system containing two centers of each type yield results that agree with the experimental data [221].

We formulated above the principal problem of molecular biophysics with respect to the cooperative properties of enzymes: determination of the nature of the active-center interaction. The presence of several equivalent active centers in an enzyme implies the presence of equivalent subunits, i.e., a quaternary structure. Direct experiments (splitting of proteins into subunits and deter-

mination of their molecular weights) showed that most ASE actually have a quaternary structure. On this basis, Monod, Wyman,
and Changeux [224] suggested a simple model of ASE. The protein
is considered as an oligomer consisting of several identical subunits (protomers), which occupy equivalent positions. The ASE
molecule should then have at least one axis of symmetry. Such a
molecule can be constructed isologously or heterologously; the
protomer-interaction region consists of two identical segments in
the first case and of two different segments in the second case
(Fig. 42). A heterologous polymer of indefinite length, as shown
in the figure, is also possible; F-actin can be considered to be an
example of such a polymer [39, 225]. The interaction of an ASE
with ligands (a substrate and ASEf) has the following characteristics:

1) each ligand corresponds to one and only one active center
in the protomer;

2) the conformation of each protomer is subject to restrictions determined by its association with other protomers;

3) the oligomer as a whole can be in two or more states differing in conformation but not in symmetry;

4) the affinity of the stereospecific centers for a ligand thus
changes when the state of the oligomer is altered.

It is readily shown that such a system is cooperative. With
no limitation of generality, we can consider a dimer that can be in
two different states, R and T. The dimer can bind 0, 1, or 2 molecules of the ligand S in each of these states. We accordingly ob-
tain the six states R_{00}, R_{10}, R_{11}, T_{00}, T_{10}, and T_{11} (two centers: a
0 indicates that the center is free and a 1 that it is occupied by an
S molecule). The equilibrium conditions have the form:

$$\left.\begin{array}{l} T_{00} = LR_{00}, \\[2mm] R_{10} = 2R_{00}\dfrac{S}{K_R}\,; \quad R_{11} = \dfrac{1}{2}R_{10}\dfrac{S}{K_R} = R_{00}\dfrac{S^2}{K_R^2}, \\[3mm] T_{10} = 2T_{00}\dfrac{S}{K_T}\,; \quad T_{11} = \dfrac{1}{2}T_{10}\dfrac{S}{K_T} = T_{00}\dfrac{S^2}{K_T^2}\,. \end{array}\right\} \qquad (113)$$

The total amount of enzyme is

$$E = R_{00} + R_{10} + R_{11} + T_{00} + T_{10} + T_{11}. \tag{114}$$

Here, K_R and K_T are the dissociation constants of S in the R and T states, while L is the equilibrium constant for the transition between states in the absence of S. The dissociation constants in states R_{10} and R_{11} (and, accordingly, T_{10} and T_{11}) are assumed to be the same. The function for saturation of the protein with substrate is written in the form

$$\overline{Y} = \frac{R_{10} + 2R_{11} + T_{10} + 2T_{11}}{2(R_{00} + R_{10} + R_{11} + T_{00} + T_{10} + T_{11})} = \frac{K_R \dfrac{1 + Lc}{1 + Lc^2} S + S^2}{K_R^2 \dfrac{1 + L}{1 + Lc^2} + 2K_R \dfrac{1 + Lc}{1 + Lc^2} S + S^2}, \tag{115}$$

where $c = K_R/K_T$. The curve for $\overline{Y}(S)$ is S-shaped and corresponds to a cooperative system. The cooperativity disappears at $c = 1$, $L \to 0$, or $L \to \infty$, and we obtain

$$\overline{Y} = \frac{S}{K_R + S}, \tag{116}$$

which corresponds to the ordinary Langmuir isotherm.

The corresponding expression for the reaction rate, obtained from the equilibrium conditions, has the form

$$v = 2Ek \frac{1 + \varkappa Lc^2}{1 + Lc^2} \frac{K_R \dfrac{1 + \varkappa Lc}{1 + \varkappa Lc^2} S + S^2}{K_R^2 \dfrac{1 + L}{1 + Lc^2} + 2K_R \dfrac{1 + Lc}{1 + Lc^2} S + S^2}, \tag{117}$$

where k is the rate constant for the states R_{10} and R_{11}, and $\varkappa k$ is the rate constant for the states T_{10} and T_{11}. At $c \ll 1$ and not overly large values of L and $\varkappa$, we find

$$\overline{Y} \approx x \frac{1 + x}{L + (1 + x)^2}, \tag{118}$$

$$v \approx 2Ekx \frac{1 + x}{L + (1 + x)^2}, \tag{119}$$

where $x = S/K_R$.

We have considered the homotropic cooperative effect inherent in the model, i.e., cooperativity with respect to a single ligand, i.e., the substrate.

Let us assume that each of the two protomers contains three active centers: one for the substrate S, one for the inhibitor I, and one for the activator A. For purposes of simplicity, we will assume that I has an affinity only for the T state of the dimer and A has an affinity only for the R state. At $c \ll 1$, the function for saturation of the enzyme with substrate in the presence of A and I has the form

$$\overline{Y} \approx \frac{1 + x}{L' + (1 + x)^2},$$
(120)

where

$$L' = L \frac{(1 + I/K_I)^2}{(1 + A/K_A)^2}.$$
(121)

It can be seen that the heterotropic effect of an ASEf affects the saturation function for S; I increases the cooperativity, while A eliminates it. The model proposed by Monod, Wyman, and Changeux is in agreement with many experimental data and permits us to make a number of predictions. It enables us to divide the effects observed into two classes, in accordance with the influence of S and the ASEf on the ASE. In the first case, the ASEf and S have different affinities for the T and R states. The ASEf alters the affinity of the protein for S and vice versa. In the second case, S has the same affinity for the two states of the ASE and the ASEf affects the reaction only when the two states of the protein differ in catalytic activity. The ASEf should exhibit cooperative homotropic interactions in the first case but not in the second. We have seen that, in conformity with the model, the cooperative kinetics of allosteric systems are governed by the shift in the R $\rightleftharpoons$ T equilibrium. This is indirect cooperativity, in contrast to the direct cooperativity considered in the preceding chapter, which reduced to the direct influence of the first ligand molecule attached on the interaction between the second ligand molecule and the protein. The saturation function and reaction rate for a dimer with direct cooperativity are expressed by Equations (97) and (98), which must be compared with Equations (115) and (117). It is obvious that the same qualitative conclusions follow from both sets of formulas, but the model with direct cooperativity is simpler and the formulas contain one less constant. The quantitative results naturally differ. Thus, for example, the value of S at half-saturation, i.e., $\overline{Y} = 0.5$, is

$$S_{0,5} = K_R \sqrt{\frac{1+L}{1+Lc^2}},$$
(122)

according to Equation (115) and

$$S_{0,5} = K\sqrt{\beta}.$$
(123)

according to Equation (98). There is as yet no way to distinguish these two types of cooperativity.

The same concepts were recently applied to biological membranes [321].

Appendix I describes the theory of graphs, which materially simplifies solution of complex problems in enzyme kinetics. Simple cooperative models of ASE with paired equilibria and both types of cooperativity have been investigated with the aid of this technique [226] and good agreement with experimental data has been obtained. As an example, let us consider the action of an ASEf that is a substrate analog. On being attached to one of the two active centers, the ASEf affects the interaction of the protomers, increasing the affinity of the unoccupied active center for S. At the same time, the ASEf occupying active centers reduces the number of centers free for S. It should thus act as an activator at low concentrations and as an inhibitor at high concentrations. It is a competitive activator. Figure 43 is a diagram of interactions involving an ASEf. The following symbols are used for the enzyme states: F_{00} is the free state, F_{0S} is the state binding one S molecule, F_{SS} is the state binding two S molecules, F_{0A} is the state binding one ASEf molecule (A), F_{AA} is the state binding two A molecules, and F_{SA} is the state with one center binding an S molecule and the other an A molecule. Let us assume that cooperativity is associated solely with the difference in the affinities of the ASE for S in different states with the same catalytic activity. Calculations by the graph method yield the following expression for the rate of S conversion:

$$v = 2k_2 k_A' ES \frac{x}{y},$$
(124)

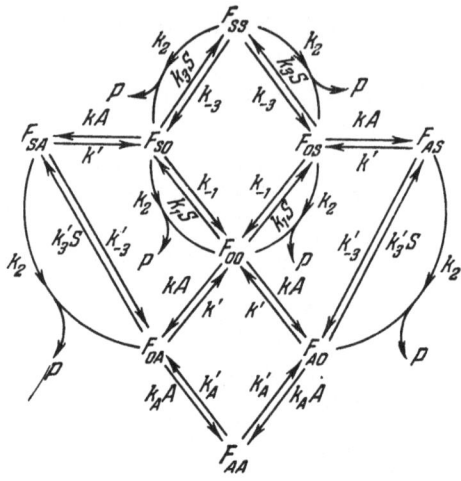

Fig. 43. Diagram of enzymatic process involving competitive activator.

where

$$x = k'k_1b'\,(a + k_1S) + abkk_3'\,A,$$

$$y = k'k_A'b\,(ab + 2bk_1S + k_1k_3S^2) + 2abkk_A'\,(b' + k_3'S)\,A + abb'kk_A A^2.$$

The meaning of the notation is clear from Fig. 43. The following
cases are possible:

1) $k_3 = k_3'$ and $b = b'$; both types of cooperativity for the substrate are present;

2) $k_1 = k_3$ and $a = b$; only indirect cooperativity for the substrate occurs;

3) $k_1 = k_3 = k_3'$ and $a = b = b'$; there is no cooperativity for
the substrate;

4) $A = 0$; only direct cooperativity occurs.

Indirect cooperativity is introduced into this model by the
presence of the ASEf.

In the first case,

$$v = 2k'_A k_2 ES \; \frac{k'k_1(b + k_3 S) + akk_3 A}{k'k'_A(ab + 2bk_1 S + k_1 k_3 S^2) + 2akk'_A(b + k_3 S)A + abkk_A A^2} \; . \tag{125}$$

At large A, we find that $v \rightarrow 0$, i.e., A acts as an inhibitor. Conversely, as A rises from zero, the curve for $v(A)$ passes through a maximum at a low value of S. An ASEf, such as maleate, acts on aspartate transcarbamylase in precisely this manner [220].

The model devised by Monod, Wyman, and Changeux and the model involving direct cooperativity are confirmed by the disappearance of cooperativity when the enzyme is split into subunits by heating or exposure to other agents. We thus obtain a third case, and v is

$$v = \frac{2k'k_1 k_2 ES}{k'(a + k_1 S) + akA} \tag{126}$$

The A acts as an ordinary competitive inhibitor. This is again in agreement with the data for aspartate transcarbamylase.

Very few structural investigations have been conducted on ASE. The nature of the changes in an ASE when it is acted upon by a substrate and ASEf has not been studied. The results of an investigation of phosphofructokinase, an ASE, were presented above (see p. 92). It was shown that an ASEf causes conformational changes in the enzyme.

Doy recently proposed a new model for the allosteric enzyme anthranylate synthetase [227]. Tryptophan is an allosteric inhibitor of this enzyme, which acts on various bacteria. In contrast to the model of Monod, Wyman, and Changeux, Doy's model is an aggregate of nonidentical subunits synthesized by different genes and performing different functions. Doy cites a number of convincing genetic proofs for his model. It is not known whether tryptophan acts on one of the subunits or only on the aggregate as a whole. It is possible that ASE include both symmetric and asymmetric oligomers and, in this sense, the theory of Monod, Wyman, and Changeux cannot be regarded as exhaustive. Hemoglobin (see Chapter 13) resembles an ASE in many respects. It contains protomers of two types, α and β, which differ in amino acid composition.

Aggregation, i.e., the occurrence of quaternary structure in ASE, is extremely important. The work of Doy [227] is of special significance in this connection, since he obtained information on aggregation from an independent source — genetic data. He emphasizes the evolutionary importance of protein aggregation as the factor responsible for development of the cell organoids. Allosteric proteins exhibit a relationship between supermolecular structure and specific regulatory function, which forms the basis for morphogenesis.

Atkinson noted that the living organism is characterized by coexistence of contradictory properties: an extreme diversity of chemical functions on one hand and very high chemical stability (homeostasis) on the other [228]. ASE play a special role in a system of this sort. Adenylates (AMP, ADP, and ATP; see p. 92) are allosteric effectors in many cases. The AMP and ATP concentrations in vital chemical systems are inversely proportional. Stimulation of an allosteric enzyme by AMP corresponds to negative metabolic feedback as a result of the inhibitory action of ATP. A situation of this type arises in the Krebs cycle.

Atkinson demonstrated that there is an interaction between the feedback effected by ASE in regulation and that effected by the operon in protein biosynthesis. The interaction of allosteric regulation and repression of induced enzyme synthesis (see [3], p.355) should be substantially more effective in producing homeostasis than allosteric regulation alone.

It is difficult to overemphasize the significance of the discovery of allosteric enzymes. Study of ASE is an important pathway to understanding of the living cell as an integral system and to determination of the molecular bases for morphogenesis and homeostasis. This area of biochemistry and molecular biophysics is still in its infancy. A whole set of extremely interesting and important problems has arisen.

The reader will find more detailed information on ASE and the broader complex of problems associated with regulation of the intracellular enzyme apparatus by metabolites in the interesting surveys by Kafiani [229].

Chapter 13

HEMOGLOBIN

Hemoglobin is not an enzyme: its function consists in transporting molecular oxygen and not in catalyzing any chemical reaction. However, it can serve as an excellent model of an enzyme, particularly an allosteric enzyme.

Hemoglobin is a protein consisting of four subunits, which normally are divided into two types, α and β. Each subunit contains a functional heme group, which is a porphyrin ring with an iron atom in the center. Heme plays the same part in hemoglobin as prosthetic groups do in enzymes. In addition, it directly serves as a prosthetic group in such enzymes as catalase, peroxidase, and the cytochromes. Oxygen (O_2) can be regarded as the substrate and other compounds that attach to heme, such as CO, can be treated as competitive inhibitors. Since hemoglobin is catalytically inactive, its properties cannot be characterized by the kinetics of any reaction other than the attachment of O_2 or other compounds. However, we can employ the saturation function as before.

This important protein has been the subject of numerous investigations, which have yielded very valuable results. Much is still unclear, but the principal molecular properties of hemoglobin, beginning with its primary structure, have been established and studied in detail. We will consider those properties that simulate the cooperative conformational properties of enzymes.

The cooperative character of the interaction of hemoglobin (Hb) with O_2 is directly shown by the shape of the curve representing saturation, or oxygenation (this term is used because we are speaking of attachment of a whole oxygen molecule rather than of

133

oxidation, i.e., of the $Fe^{2+} \to Fe^{3+}$ transition). The dependence of the degree of oxygenation on the partial oxygen pressure p is characterized by an S curve resembling that in Fig. 36. This function is expressed mathematically by the Hill equation

$$\overline{Y} = \frac{p^{\alpha}}{K + p^{\alpha}}, \tag{127}$$

where $\alpha \approx 2.8$. The superscript α cannot be assumed to equal the number of interacting subunits n; α is some interaction coefficient associated with n. If we generalize Equation (115) into a system for n rather than two protomers, we obtain [224]

$$\overline{Y} = \frac{Lcx(1 + cx)^{n-1} + x(1 + x)^{n-1}}{L(1 + cx)^{n} + (1 + x)^{n}}, \tag{128}$$

where $x = S/K_R$. For equine hemoglobin, n = 4, L = 9054, and c = 0.014. The presence of an inflection in the curve for $\overline{Y}(p)$ expresses the correspondence between the structure and biological function of Hb. Oxygenation increases sharply when a definite partial oxygen pressure is reached, in conformity with the homeostatic regime of the organism. In order to interpret the properties of Hb, they must be compared with those of another complex protein, myoglobin (Mb). The function of Mb is to store oxygen for subsequent participation in oxidative phosphorylation in the muscles. Mb contains one heme group and consists of a single globule resembling an Hb protomer. It has no quaternary structure. The dependence of $\overline{Y}$ on p is thus noncooperative; the saturation curve is simple and resembles the Langmuir isotherm:

$$\overline{Y} = \frac{p}{K' + p}, \tag{129}$$

or, substituting n = 1 into Equation (128),

$$\overline{Y} = \frac{x}{(L+1)(Lc+1)^{-1} + x} = \frac{p}{\varkappa(L+1)(Lc+1)^{-1} + p}, \tag{130}$$

where K' and $\varkappa$ are constants.

The cooperative properties of Hb are thus determined by its quaternary structure. We can calculate Y by treating the cooperativity as a direct interaction between the protomers. As has already been pointed out, Hb consists of two protomers of types α and β. Let us designate the relative free energies of the interaction of the corresponding protomers as $F_{xy}^{(1)}$ (where x and y are α and β) if protomer x is occupied by O_2 and protomer y is free and as $F_{xy}^{(2)}$ if both protomers are occupied. The free energies of the states HbO_2, HbO_4, HbO_6, and HbO_8 should thus obviously contain the terms $kT \log p$, $2kT \log p$, $3kT \log p$, and $4kT \log p$. We will employ the notation:

$$a_1 = \exp\left(\frac{F_{\alpha\alpha}^{(1)}}{kT}\right), \quad b_1 = \exp\left(\frac{F_{\alpha\beta}^{(1)}}{kT}\right), \quad c_1 = \exp\left(\frac{F_{\beta\beta}^{(1)}}{kT}\right),$$

$$a_2 = \exp\left(\frac{F_{\alpha\alpha}^{(2)}}{kT}\right), \quad b_2 = \exp\left(\frac{F_{\alpha\beta}^{(2)}}{kT}\right), \quad c_2 = \exp\left(\frac{F_{\beta\beta}^{(2)}}{kT}\right).$$

The results obtained are presented in Table 14. We have

$$\bar{Y} = \frac{1}{4}\frac{\partial \ln Z'}{\partial \ln p} = \frac{Z'}{4Z'} =$$

$$= \frac{2(a_1 + c_1)b_1^2 p + 2(a_2 b_1^2 + 4a_1 c_1 b_2 + b_1^2 c_2)b_1^2 p^2 + 6(a_1 c_2 + a_2 c_1)b_1^2 b_2^2 p^3 + 4a_2 b_2^4 c_2 p^4}{4[1 + 2(a_1 + c_1)b_1^2 p + (a_2 b_1^2 + 4a_1 c_1 b_2 + b_1^2 c_2)b_1^2 p^2 + 2(a_1 c_2 + a_2 c_1)b_1^2 b_2^2 p^3 + a_2 b_2^4 c_2 p^4]}$$

$$\tag{131}$$

Comparison of Equations (131) and (127) shows that α is actually not identical to n. For simplicity, let us assume that $a_1 = b_1 = c_1 = a$ and $a_2 = b_2 = c_2 = b$. Then,

$$\bar{Y} = a^3 p \frac{1 + 3abp + 3b^3 p^2 + b^6 p^3}{1 + 4a^3 p + 6a^4 bp^2 + 4a^3 b^3 p^3 + b^6 p^4}. \tag{132}$$

The cooperativity disappears at $b = a^2$. In this case,

$$\bar{Y} = \frac{a^3 p}{1 + a^3 p}. \tag{132a}$$

It must be assumed that $b = fa^2$, where $f > 1$. Introducing the notation $a^3 p = p'$, we obtain

$$\bar{Y} = \frac{p' + 3fp'^2 + 3f^3 p'^3 + f^6 p'^4}{1 + 4p' + 6fp'^2 + 4f^3 p'^3 + f^6 p'^4}. \tag{132b}$$

At certain values of f and A, Equation (132b) should be identical to Equation (127). The first such evaluation of the interaction of the Hb subunits, calculating Z and $\bar{Y}$, was made by Pauling [229] (see also [230]). The phenomenological thermodynamics of systems cooperatively interacting with ligands in the manner of Hb was worked out by Wyman [232, 233]. This theory establishes a relationship among the interactions of a protein (oligomer) with different ligands

Table 14. Calculation of Equilibrium in Hb + O₂ System

Type of molecule	Chains occupied	Statistic. wt.	Relative free energy	Contribution to Z	No. of O₂ molecules attached	Contribution to Z'
Hb	—	1	0	1	0	0
HbO₂	α	2	$-2F^{(1)}_{\alpha\beta} - F^{(1)}_{\alpha\alpha} - kT\ln p$	$2a_1b_1^2 p$	1	$2a_1b_1^2 p$
HbO₃	β	2	$-2F^{(1)}_{\alpha\beta} - F^{(1)}_{\beta\beta} - kT\ln p$	$2b_1^2 c_1 p$	1	$2b_1^2 c_1 p$
HbO₄	αα	1	$-4F^{(1)}_{\alpha\beta} - F^{(2)}_{\alpha\alpha} - 2kT\ln p$	$a_2 b_1^4 p^2$	2	$2a_2 b_1^4 p^2$
HbO₄	αβ	4	$-F^{(2)}_{\alpha\beta} - 2F^{(1)}_{\alpha\beta} - F^{(1)}_{\alpha\alpha} - F^{(1)}_{\beta\beta} - 2kT\ln p$	$4a_1 b_1^2 c_1 b_2 p^2$	2	$8a_1 b_1^2 c_1 b_2 p^2$
HbO₄	ββ	1	$-4F^{(1)}_{\alpha\beta} - F^{(2)}_{\beta\beta} - 2kT\ln p$	$b_1^4 c_2 p^2$	2	$2b_1^4 c_2 p^2$
HbO₆	ααβ	2	$-2F^{(2)}_{\alpha,?} - 2F^{(1)}_{\alpha\beta} - F^{(1)}_{\alpha\alpha} - F^{(1)}_{\beta\beta} - 3kT\ln p$	$2a_2 b_2^2 c_1 p^3$	3	$6a_2 b_1^3 b_2^2 c_1 p^3$
HbO₆	αββ	2	$-2F^{(2)}_{\alpha\beta} - 2F^{(1)}_{\alpha\beta} - F^{(2)}_{\beta\beta} - F^{(1)}_{\alpha\alpha} - 3kT\ln p$	$2a_1 b_2^3 c_2 p^3$	3	$6a_1 b_1^3 b_2^2 c_2 p^3$
HbO₈	ααββ	1	$-4F^{(2)}_{\alpha\beta} - F^{(2)}_{\alpha\alpha} - F^{(2)}_{\beta\beta} - 4kT\ln p$	$a_2 b_2^4 c_2 p^4$	4	$4a_2 b_2^4 c_2 p^4$

and formally accounts for a number of the properties of Hb. Wyman's theory is essentially equivalent to the later theory of indirect cooperativity in allosteric enzymes, which we considered above.

Attachment of O_2 to Hb is actually accompanied by effects that, in aggregate, can be attributed to conformational changes. Let us enumerate these effects [234]:

1) Hb and HbO_8 have different crystal structures and solubilities.

2) If rapid dissociation of $Hb(CO)_8$ is induced by flash photolysis, the Hb formed reacts more rapidly with O_2 or CO than ordinary Hb [235].

3) The reactivities of HbO_8 and Hb differ. Thus, certain phospholipids split the hemoglobin bond more rapidly in HbO_8 than in Hb.

4) The dye bromthymol blue has a substantially greater affinity for Hb than for HbO_8, which is reflected in the reaction kinetics. Large amounts of this dye, up to 10 molecules per heme unit, can be attached without saturation. Hemoglobin that has bound the dye has a lower affinity for oxygen, although the dye is attached to the globin and not the heme. In this sense, the effect is very similar to the reactions of allosteric enzymes. It depends to a large extent on the medium pH; the pK of the dye changes from 7.1 to 8 when it is bound by the protein [236].

5) The reactivity of the SH groups with respect to n–methylmaleimide and particularly iodacetamide changes when Hb is oxygenated. HbO_8 is more reactive than Hb. This effect does not occur in Mb or in MbH, which consists of four identical globules of the β type [237].

6) The kinetics of alkaline denaturation change when Hb is oxygenated.

Finally, there is the Bohr effect. This interesting phenomenon has been the subject of many investigations. In essence, it consists in an increase in the degree of dissociation of the imidazole rings of the histidine residues when oxygen is attached to the heme. Three moles of H^+ are split off per mole of Hb. The Bohr effect thus demonstrates the relationship between the interactions of Hb with O_2 and with H^+. It can be measured either from the change in H^+ binding during oxygenation by differential titration or from the change in O_2 binding when the pH is varied.

The results obtained are well described by Wyman's formal theory [238]. However, the molecular nature of the effect cannot be considered fully resolved, despite the interesting works devoted to research on this problem (see [234-239]). It might be surmised that we are dealing with an example of the trans-influence discovered by Chernyaev. This effect is observed in complex compounds of transition-metal atoms with coordination numbers of four or six; it is manifested in the fact that, when one of the ligands in the coordination sphere is replaced, there is a change in the interaction between the central atom and the ligand on the opposite side, in the trans-position with respect to the substituted group, i.e., on the diagonal of a square or octahedron (see [240]). The Fe^{++} in hemoglobin forms an octahedral complex: the four coordination bonds in the plane of the heme porphyrin ring lead to the nitrogen atoms of pyrrole groups, one bond connects the Fe^{++} to the imidazole ring of a histidine residue, and a co-linear bond is directed outward and leads to H_2O in the case of Hb and to O_2 in the case of HbO_8 (see Fig. 3). The heme is bonded to the globin in both Hb and HbO_8; replacement of the ligand, i.e., substitution of oxygen for water, occurs during oxygenation. The complex is octahedral in both Hb and HbO_8. However, such ligand replacement causes detachment of protons from histidine residues other than that attached to the Fe^{++}. The Bohr effect cannot be attributed directly to the trans-influence, since it is not observed during oxygenation of Mb, in which the heme is attached to the same histidine. It is apparently associated with conformational transformations of Hb and characterizes its allosteric properties.

At the same time, there is a trans-influence in heme and it is of great importance for the properties of heme-containing proteins. Urry and Eyring explained electron transport in cytochromes from the theory of the trans-influence (the "imidazole pump" model) [314]. Atanasov investigated the conformational properties of myoglobin having different ligands and found that the conformational stability of this protein varies regularly when the ligand in the heme group is changed [315]. Investigation of the same myoglobin derivatives by the magnetic-rotation method (see Appendix II) showed that the electronic state of the prosthetic group affects the properties of the protein as a whole.

In connection with the foregoing, we can formulate another general problem of enzyme physics: the influence of the electronic state of the prosthetic group, coenzyme, or cofactor (particularly transition-metal ions) on the conformational properties of an enzyme, which govern its activity. We are not speaking of any special semiconductive properties of proteins, which are dielectrics, but of specific interactions of the trans-influence type in which such amino acid residues as His and other aromatic residues participate.

Rossi-Fanelli, Antonini, and Caputo[241] discovered an interesting effect. Both Hb and HbO_8 dissociate into subunits in concentrated salt solutions. However, $x = 2.7$-2.8 under these conditions, falling to 1.4 at low salt concentrations. It is clear from the calculations made above that the S shape of the curve for $\overline{Y}(p)$ should result from protomer interactions; the value of x should consequently be of the order of one for Hb (and Mb) in the dissociated state.

This paradox can be explained only by combination of the subunits during oxygenation and subsequent dissociation. It is possible that a dissociation equilibrium related to the oxygen is realized; this would be a special limiting case of a conformational change, i.e., a change in quaternary structure [232]. It has actually been established that oxygenation promotes dissociation of hemoglobin into dimers [242].

It was recently demonstrated that Hb molecules are capable of subunit exchange [243]. Hb and HbO_8 contain ferrohemes with Fe^{++} (iron atoms in the divalent state). An equimolar mixture of solutions of hemoglobin and methemoglobin (oxidized Hb containing ferriheme, Fe^{+++}) has a greater affinity for O_2 than ferrohemoglobin. This is due to the fact that subunit exchange occurs between the ferri and ferro forms. Benesch et al. [243] considered the following reactions:

1. Oxygenation of the ferro form (low affinity for O_2):

$$\frac{\alpha\beta}{\beta\alpha} \rightarrow \frac{\alpha_{O_2}\beta_{O_2}}{\beta_{O_2}\alpha_{O_2}}.$$

2. Dissociation of HbO_8:

$$\beta_{O_2}\alpha_{O_2}^{\alpha_{O_2}\beta_{O_2}} \rightarrow 2\alpha_{O_2}\beta_{O_2}.$$

3. Exchange with the ferri form (designated by α^+ and β^+):

$$\beta^+\alpha^+ + \alpha_{O_2}\beta_{O_2}^{\alpha^+\beta^+} \rightarrow \beta_{O_2}\alpha_{O_2}^{\alpha^+\beta^+} + \alpha^+\beta^+.$$

4. Oxygenation of the mixed ferro—ferri form (high affinity for O_2):

$$\beta_{O_2}\alpha_{O_2}^{\alpha^+\beta^+} \rightarrow \beta_{O_2}\alpha_{O_2}^{\alpha_{O_2}\beta_{O_2}}.$$

These authors obtained data indicating that such a process is possible [243]. They deny that transhemination, i.e., transfer of a heme group from protein to protein, which was studied in detail by Blyumenfel'd [244], can participate in it, since this transfer takes place comparatively slowly. A new hypothesis of oxygenation through exchange between the ferri and ferro forms has been advanced. The S shape of the curve for $\overline{Y}(p)$ results from progressive replacement of reaction 1 by reaction 4; there is an increase in affinity for oxygen as oxygenation proceeds. This hypothesis requires more detailed confirmation. If it is valid and oxygenation is actually coupled with oxidation, i.e., with the $Fe^{2+} \rightarrow Fe^{3+}$ transition, it will materially change our notions of the properties of hemoglobin. In a recent article, the same authors attempted to explain the allosteric properties of Hb as resulting from subunit exchange and hence from the dissociation equilibria considered above [316].

Hemoglobin is an excellent model of an enzyme both because it is comparatively easy to study its interaction with oxygen and other ligands, since they do not undergo chemical transformations, in contrast to the substrates of enzymatic reactions, and because it is possible to conduct detailed investigations of its structure (and that of myoglobin) by the direct technique of x-ray diffraction analysis, as has already been done in the brilliant work of Perutz and Kendrew (see [10, 11, 38]).

As was pointed out above (see p. 28), the amino acid residues within the Hb and Mb globules are usually nonpolar. Com-

Table 15. Distances Between
Iron Atoms of Four Hemes
(in Å)

	Hb, human	HbO$_8$, equine
Fe$_{\beta1}$ — Fe$_{\beta2}$	40.3	33.4
Fe$_{\alpha1}$ — Fe$_{\alpha2}$	36.0	35.0
Fe$_{\beta1}$ — Fe$_{\alpha1}$	25.2	25.0
Fe$_{\beta1}$ — Fe$_{\alpha2}$	30.4	37.4

parison of the amino acid sequences of different hemoglobins and myoglobins shows that the presence of even one group with a large dipole moment makes the tertiary structure of the globule unstable [38]. Almost all polar residues are in contact with water, either at the surface of the molecule or in the internal interglobular cavity of hemoglobin. These important facts were established by x-ray diffraction analysis.

Muirhead and Perutz [246] used the same technique to demonstrate that attachment of oxygen to hemoglobin causes a substantial change in its quaternary structure. Table 15 gives the distances between the iron atoms of the four hemes in Hb and HbO$_8$.

There are strong arguments to support the thesis that the hemoglobins of man and the horse are almost identical. Table 15 indicates that oxygenation is accompanied by a decrease in the distances between the iron atoms of the hemes: the hemoglobin molecule "breathes," binding and releasing oxygen.

The approximation of the globules during attachment of the "substrate" (oxygen) should lead to an intensification of their interaction and specifically to greater splitting of the resonance frequencies which, as we have seen, is in agreement with the illustrative calculation made on p. 116. This however cannot be considered a proof of the resonance nature of the cooperative interaction in hemoglobin. However, it would be sensible to investigate this possible explanation for the specific properties of hemoglobin, something that has not yet been done.

The difference in the molecular structures of Hb and HbO$_8$ is directly manifested in a difference in the structures of their crystals. The resolving power of x-ray diffraction analysis (5.5 Å in the investigation conducted by Muirhead and Perutz [256]) is insufficient to reveal changes in the secondary structure of hemoglobin during oxygenation. Such changes do not necessarily accompany changes in quaternary structure. Actually, study of the dispersion of optical activity has shown that the α-helicity of hemoglobin remains constant during oxygenation, within the limits of experimental error [247, 248]. It does not follow from this that the tertiary structure is unaltered; the ordered portion of the polypeptide chain can be dehelicized and the disordered portion helicized during oxygenation, still retaining roughly the same total proportion of α-helix. However, this is to some extent refuted by the considerations advanced by Guzzo (see p. 16). Investigation of the anomalous dispersion of optical rotation for hemoglobin and myoglobin in the natural absorption bands of the heme group has yielded interesting results (see [249], in which the present author and his colleagues have clarified their previous data [250]). The difference in the AORD curves for Hb and Mb on the one hand, and HbO_8 and MbO_2 on the other is very marked, but the greatest difference is in the trends of the AORD curves for HbO_8 and MbO_2, which indicates interaction of HbO_8 subunits similar to MbO_2. These data naturally require detailed interpretation.

Even more striking are the results of an investigation of the anomalous dispersion of magnetic rotation for Hb, Mb, HbO_8, and MbO_2; these are described in Appendix II.

The most pressing problem in the physics of hemoglobin is determination of the mechanism of the cooperative interactions. The physicochemical and structural results discussed above have not provided a solution to this problem, although they have yielded valuable information toward a solution.

Study of hemoglobin is very important both in connection with its specific biological function and because it is similar to allosteric enzymes. Determination of the character of the internal interactions in hemoglobin is vital for enzyme physics and for protein physics as a whole.

Chapter 14

ENZYMES AND MECHANOCHEMISTRY

All the foregoing material enables us to advance the following general hypothesis: the action of enzymes is closely associated with their macromolecular conformational properties and, in this sense, the factors underlying enzymatic activity are related to those underlying the high elasticity of rubber-like polymers.

A change in the secondary, tertiary, or quaternary structure of an enzyme macromolecule implies spatial displacement of atoms or groups of atoms present in it, i.e., mechanical movement. This obviously produces mechanical work, whose source is the free energy liberated during the enzymatic reaction. We can therefore treat the conformational transformation of an enzyme, particularly its achievement of structural correspondence with a substrate, as a mechanochemical process, i.e., one in which chemical energy is directly converted to mechanical work. Hence, there naturally arises the concept of the "rack" [74], i.e., deformation of a substrate molecule interacting with the active center of an enzyme. The mechanical energy can in turn be used for chemical conversion of the substrate.

This interpretation yields little that is new so long as we limit ourselves to enzymatic reactions in solution. In final analysis, relative displacement of atoms (nuclei) occurs in any chemical reaction, but this does not give us grounds for speaking of mechanochemical phenomena; it is irrational to distinguish a mechanical form of atomic movement in this situation. We can speak of mechanochemistry with respect to an enzyme, since it is a macrosystem containing many atoms and executing mechanical movement as a whole. However, this is merely substituting the term "mechanical" for "conformational."

143

On the other hand, if an enzyme molecule is incorporated into a macroscopic supermolecular structure, the latter can, when
properly organized, carry out mechanical movement as a whole
and produce mechanical work through the free energy of a chemical enzymatic reaction. Some or all of the working compounds of
mechanochemical systems therefore can and should be enzymes.
Actually, all intracellular movements and movements of cells and
of the organism as a whole are carried out under isothermal and
isobaric conditions. They result from conversion of chemical (in
the broad sense of the word) free energy to mechanical work and
are mechanochemical processes. However, all biochemical reactions involve the participation of enzymes. The mechanochemistry
of living systems is therefore enzyme mechanochemistry.

Let us enumerate the principal mechanochemical processes
encountered in biology:

1. Movement of animals, or muscular work.

2. Movement of plants.

3. Movement of cells, or the work of flagella and cilia.

4. The entire aggregate of movements during mitosis and
meiosis.

5. The movement of protoplasm within the interphase cell.

6. Mechanochemical phenomena in membranes.

7. The movement of mRNA relative to the ribosomes in the
polysomes during protein synthesis.

8. The contractile processes in phage tails.

This list is not exhaustive and it can be assumed that further
research will demonstrate the mechanochemical nature of other
biological processes, particularly those taking place within the cell.

It has been established that, in the overwhelming majority of
instances, the energy balance of a biological mechanochemical system is governed by the free energy liberated during cleavage of
ATP (adenosine triphosphate). Such cleavage requires the participation of ATPase (the corresponding enzyme or enzymes). The
working compounds of most mechanochemical processes are contractile proteins. One of the most important problems in this area

of molecular biology and biochemistry is determination of the relationship between the mechanical properties of a contractile system and ATPase activity.

Modern theories of the properties of contractile proteins are based on a discovery made by Engelhardt and Lyubimova, who established that myosin (the contractile protein of muscle) is an ATPase [251, 252]. Thus, the contractile protein that serves as the working compound in a mechanochemical system performing mechanical work and the enzyme responsible for the chemical reaction in which the energy converted to work is liberated proved to be the same protein. Engelhardt formulated the general principle that can be regarded as the key to the problem as a whole: "The enzyme that catalyzes a decisive biochemical reaction in functional metabolism should be an integral part of the mechanism that effects the function in question" [253].

Current concepts of muscular activity proceed from the conformational transformations of a complex muscle protein, actomyosin ([3], Chapter 9). In one of the most important theories of muscle contraction, that of Davies, the relative displacement of actin and myosin fibrils, which governs this process, is related to the ATPase activity of myosin, which undergoes conformational transformations during ATP dephosphorylation and the reverse process, i.e., ADP phosphorylation [3, 254].

A similar model was independently proposed by Tonomura et al. [255]. Let us quote the concluding words of their article: "In performing its physiological function, a protein acts mainly either as a structural unit for transformation of energy or as a metabolic catalyst. We suggest that a protein be called a 'transconformer' when its physiological function is effected by conformational rearrangement, in the same manner as a protein is called an enzyme when its function is catalysis. All energy conversions in biological systems can then be regarded as resulting from the action of a protein as a 'transconformer,' an action accompanied by liberation of energy in ATP hydrolysis."

However, an enzyme is also essentially a "transconformer" and there is no reason to distinguish a "physiological function effected by conformational rearrangement" from a catalytic, enzymatic function. The entire foregoing discussion has demonstrated that catalysis is effected through conformational transformations.

Current theories of muscular activity, which are far from conclusive, have been discussed elsewhere [3]. We will here consider only one important feature of a "transconformer," i.e., a contractile protein or enzyme.

All our concepts of the role of conformation in enzymatic activity lead to the conclusion that a forcible change in enzyme conformation induced by a mechanical agent should cause a change in enzymatic activity. This hypothesis, which I first formulated in 1962 [256], was recently confirmed experimentally. Vorob'ev and Kukhareva subjected a myosin macromolecule to deformation in a hydrodynamic field (in a dynamooptimeter). Deformation caused a change in the ATPase activity of the protein [257]. This result can be regarded as a new proof of the role of conformational changes in enzymatic catalysis.

Most of the works that have considered the properties of mechanochemical contractile systems have naturally been devoted to muscular activity. However, much has also been learned in investigations of other biological systems in which mechanochemical processes take place. Poglazov has given a general survey of the structure and properties of contractile proteins [258]. Here we will discuss only a few data pertaining to the functions of such contractile systems as flagella and membranes.

A flagellum is a supermolecular fibrillar structure that executes undulatory movements. It has been demonstrated that the working compound of the system is a contractile protein similar to actomyosin (AM) and having ATPase activity. Silvester and Holwill proposed a molecular model of flagellar function based on conformational concepts [259]. A flagellum consists of nine peripheral fibrils and two central fibrils. The outer fibrils are capable of longitudinal contraction in localized areas. The contraction cycle begins when mechanical deformation induced in a given area of a fibril by an incoming contractile wave alters the steric and electrostatic environment and hence its local conformation. The change in conformation leads to cleavage of ATP. The energy liberated permits the fibril to alter its conformation and thus perform work. Without going into further detail, we can see that the mechanical process is related to a conformational transformation of the protein, which governs its enzymatic activity.

The results of research on mitochondria are exceptionally interesting. Oxidative phosphorylation of ADP to ATP, which is coupled with a flow of electrons maintained by the system of oxidation—reduction enzymes (cytochromes), occurs in these cell organoids. The necessary ions are transported to the mitochondria from the ambient medium, a process that takes place through a membrane. It has been established that the membrane, which has a complex structure, exhibits contractile properties [260]. Extension or contraction of the membrane regulates its permeability. This is apparently true of all cell membranes. As before, the energy source is ATP. It follows from the foregoing that a membrane should contain an enzyme with ATPase activity. Venkstern and Engelhardt predicted the existence of such an enzyme in erythrocyte membranes [261]; this prediction was subsequently confirmed. Neifakh and his colleagues isolated a contractile protein, an enzyme whose properties are virtually identical to those of AM, from mitochondrial membranes [262-265]. Neifakh established that the function of mitochondria is associated with their ability to release a special phospholipid (kinasin) that allosterically activates phosphoglycerate kinase release into the medium. The contractile protein (AM) serves a catalytic function in membrane transport, as is proved by the formation of a complex of AM with kinasin. The transport process requires energy, which is supplied by ATP; decomposition of the complex is accompanied by dephosphorylation of ATP. Neifakh concluded that reversible conformational changes induced in AM by ATP are responsible for these processes.

Proceeding from these and similar phenomena, Green formulated general hypotheses for any cell membrane [266]. All membranes should have the same set of principal proteins, specifically containing a contractile protein with ATPase activity (inherent in all membranes), which is responsible for the ability to contract characteristic of all membranes. The contractile protein, or ATPase, occupies the key position in all energy conversions.

There is every reason to assume that conformational transformations of proteins govern their most diverse functions: enzymatic, mechanochemical, oxygen transfer by hemoglobin, etc. The concepts of macromolecular physics thus enable us to lay out a general approach to research on the properties of proteins, particularly enzymes.

Biological contractile systems are self-regulating. In other words, feedback between mechanical and chemical processes should occur in muscle fibers and mitochondria. Actomyosin and similar proteins have a quaternary structure; actin, a constituent of AM, polymerizes in the helical F-form ([3], p. 451). This leads us to the notion that allosteric phenomena may take part in biological mechanochemical processes. It must be emphasized that mechanochemistry has still not been considered from these standpoints. It is possible that determination and analysis of the allosteric properties of contractile proteins, which are enzymes, will prove helpful in understanding the functioning of muscles, mitochondria, flagella, etc. Thus, for example, there is reason to assume that an allosteric mechanism participates in protein synthesis, in the displacement of ribosomes along the mRNA in a polysome [267, 268] (see also p. 37).

It is thought that conformational transformations of a mechanochemical nature, which are perhaps also associated with allosterism, play some role in the molecular model of the "sodium pump" (see [269]). The propagation of a neural impulse along an axon results from a change in the permeability of the axon membrane to Na^+ and K^+ ions. Opit and Charnock [270] have given a graphic description of this process, starting from conformational transformation of the ATPase present in the axon membrane.

* * * * * * * * * * * *

This book has been devoted to enzyme physics. At the current level of research, this title may sound too grand. Enzyme physics is essentially unformulated and our knowledge in this area is very vague. However, the extensive variety of data that have been amassed indicates that the conformational properties of enzymes are very important. They must be investigated by the methods of experimental and theoretical physics.

In speaking of conformations, we have in mind the higher structural levels of proteins, which are complex macromolecular systems.

The conformation aspect of enzymatic activity is, of course, inseparably associated with the chemical aspect. It would be not only erroneous but actually detrimental to science if we were to oppose physics to chemistry. Progress in enzymology requires a combination of physical and chemical concepts.

Appendix I
KINETICS OF
ENZYMATIC REACTIONS AND
GRAPH THEORY

Kinetic problems in enzymology can be very complex, since an enzyme molecule often has many active centers with respect to substrates and other effectors. Certain cases of this type were described in Chapters 11 and 12. The number of kinetic constants characterizing a steady-state enzymatic process obviously increases rapidly as the system becomes more complicated, and it is therefore hard to count on determining them independently. However, the general expression for reaction rate as a function of the concentrations of the substrate, inhibitor, etc., yields valuable information on the nature of the process, even if the individual kinetic constants are unknown. Thus, the presence of an inflection in the curve for $v(S)$ directly indicates that the process is cooperative, regardless of the quantitative dependence of v on S (see p. 113). In complex cases of steady-state kinetics, the explicit form of the function $v(S, I, \ldots)$ must be obtained by solving a system of equations in algebraic rather than numerical form. This cumbersome and tedious job reduces to evaluation of determinants. Various simplified methods have been proposed, among which is that of King and Altman [271, 272]. A classification of enzymatic reactions was subsequently developed on the basis of their method [273]. However, this technique is very complex when applied to a sufficiently large number of simultaneous kinetic equations. Solution of such problems can be materially simplified with the aid of graph theory [274]. This theory is a branch of topology and is widely employed in calculations for electrical circuits [275-279], in cybernetics, etc. Graphs are applicable to any problem pertaining to ramified flows of charges, materials, or information. The graph-theory method was first applied to enzyme kinetics by Vol'kenshtein and Gol'dshtein [280].

Let us assume that an enzyme has several active centers with respect to substrate S, inhibitor I, activator A, etc. Some of these centers may coincide. Under steady-state conditions, the system will contain enzyme molecules with different occupied active centers. We will designate the concentrations of such molecules as $F_{ijk...}$, where i, j, k, . . . can have the values 0, 1, 2, . . .; a 0 designates a free center, a 1 a center occupied by an S molecule, a 2 a center occupied by an I molecule, etc.

Each state $F_{ijk...}$ can be obtained from other states either by attachment of S, I, A, . . . to a free active center or by release of a previously occupied active center. It is convenient to employ the notation $F_{ijk...} = F_r$, where the single subscript r = 1, 2, . . . , n enumerates all states possible for the system in question, whose total number is n. The reaction rate, i.e., the rate of irreversible product formation, is written in the form

$$v = \dot{P} = \sum_{r=1}^{n} k_r F_r, \tag{I.1}$$

where k_r is the rate constant and P is the product concentration. If a reaction product is not formed in a given state F_r, the corresponding constant k_r is zero.

If steady-state conditions obtain, i.e., if all $\dot{F}_r = 0$, we obtain the following homogeneous system of n − 1 concentration equations:

$$F_t \sum_{r=1}^{n} a_{tr} = \sum_{s=1}^{n} a_{st} F_s, \quad t = 1, 2, \ldots, n - 1. \tag{I.2}$$

The coefficients a_{tr} and a_{st} contain the rate constants and the corresponding concentrations of the ligands S, I, A, . . . System (I.2) represents the equilibrium for each state F_t. Examples of such equation systems were given on pp. 112 and 122. In addition to (I.2), it is necessary to introduce the condition that the sum of the concentrations of all the enzyme states be constant, i.e.,

$$\sum_{r=1}^{n} F_r = E. \qquad (I.3)$$

The number of independent variables is thus $n - 1$ and system (I.2) has a unique solution. Having found all the F_r, we calculate v from Equation (I.1).

The enzymatic process is readily depicted by connecting the states F_r with arrows indicating the directions of the individual reactions. Each arrow has a "label" characterizing the conversion rate. Such a diagram is shown in Fig. 37.

In topology, diagrams of this sort are called linear graphs. The graph is equivalent to Equations (I.2) and its solution is the solution of the system. The fact that the solution of the graph can be found from simple rules, without special calculations, is important. A graph is a diagrammatic representation of the relationships between certain variables (F_r), which are called nodal values. The different states are represented by points, which are called vertices. The arrow connecting the two vertices r and s is called the branch $r \rightarrow s$; the value of the branch is the corresponding coefficient a_{rs}. The term path refers to a continuous sequence of branches in some one direction, with none of the vertices being encountered more than once. The value of the path is the product of the branches it incorporates.

One vertex can be chosen to be the starting point (base). The base tree is the aggregate of all branches passing through all the vertices and directed toward the base. The branches of the base tree do not form loops.

Finally, the sum of the values of all the base trees directed toward a given base r is designated as the base determinant D_r of the graph. Using Mason's rule for graphs representing electrical circuits [275], we obtain an equation for the rate of the enzymatic reaction in the form

$$v = \frac{E \sum_r k_r D_r}{\sum_r D_r}. \qquad (I.4)$$

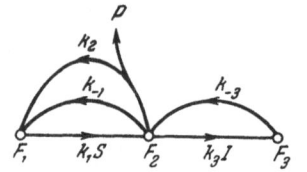

Fig. 44. Graph for noncompetitive
inhibition.

Calculation of the rate therefore reduces to finding all the base determinants D_r. Solution of this problem
is materially simplified by the graph
properties listed below.

a) "Parallel" branches are ad-
ditive. In other words, several
branches running from vertex m to
vertex r can be replaced by a single branch with a value equal to
the sum of the values of the separate branches:

$$a_{mr} = a_{mr}^{(1)} + a_{mr}^{(2)} + \cdots \qquad (I.5)$$

b) The graph can be materially simplified if, making use of
its symmetry, certain of its branches can be combined.

c) If the number of vertices is large, it is difficult to find
all the trees of the graph. The order of a graph (the number of
vertices) can be reduced. The base r and the auxiliary vertex m
are used to calculate the determinant D_r. The choice of an auxiliary vertex is arbitrary and is dictated by the structure of the
graph. Let us consider all the paths from m to r, which have the
values $P_{mr}^{(1)}$, $P_{mr}^{(2)}$, ... If each path is compressed to a point, we
obtain a lower-order graph. The base of this graph is a composite
vertex. This technique can be used to obtain a new lower-order
graph for each path, with the base determinants $D_{mr}^{(1)}$, $D_{mr}^{(2)}$, ... ,
respectively. Then,

$$D_r = P_{mr}^{(1)} D_{mr}^{(1)} + P_{mr}^{(2)} D_{mr}^{(2)} + \cdots \qquad (I.6)$$

In the case of complex graphs, the procedure for reducing their
order is repeated until the remaining graph has been converted to
a single branch.

d) If a graph contains several segments having a common
vertex, each base determinant equals the product of the base determinants of these segments. The branches in each segment of the
graph must be directed toward a common base.

We will clarify the foregoing by solving three simple problems.

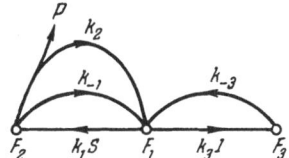

Fig. 45. Graph for competitive
inhibition.

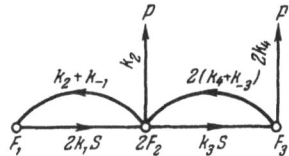

Fig. 46. Simplified graph for cooperative
system containing two equivalent active
centers.

In noncompetitive inhibition, we are dealing with three enzyme states: $F_1 = F_{00}$ is the free enzyme, $F_2 = F_{10}$ is the enzyme with substrate S attached (a Michaelis complex), and $F_3 = F_{11}$ is the enzyme with substrate S and inhibitor I attached.

The corresponding graph is shown in Fig. 44. Adding the "parallel" branches, we obtain $a_{21} = k_2 + k_{-1}$. All the base determinants then contain one tree each. Each base determinant can be regarded as the product of the base determinants of the two segments of the graph having a common vertex. We obtain

$$D_1 = (k_2 + k_{-1})\, k_{-3}, \quad D_2 = k_1 S k_{-3}, \quad D_3 = k_1 S k_3 I$$

and, in accordance with Equation (I.4),

$$v = E\,\frac{k_2 D_2}{D_1 + D_2 + D_3} = E\,\frac{k_2 k_1 S k_{-3}}{(k_2 + k_{-1})\, k_{-3} + k_1 S k_{-3} + k_I S k_3 I} = \frac{k_2 E S}{K + S + K_I S I}, \quad \text{(I.7)}$$

where

$$K = \frac{k_1 + k_2}{k_1}, \quad K_I = \frac{k_3}{k_{-3}}.$$

The linear graph has a different form for competitive inhibition (Fig. 45). We obtain

$$D_1 = (k_2 + k_{-1})\, k_{-3}, \quad D_2 = k_1 S k_{-3}, \quad D_3 = (k_2 + k_{-1})\, k_3 I$$

and

$$v = E\,\frac{k_2 k_1 S k_{-3}}{(k_2 + k_{-1})\, k_{-3} + k_1 S k_{-3} + (k_2 + k_{-1})\, k_3 I} = \frac{k_2 S E}{K + S + K K_I I}. \quad \text{(I.8)}$$

Finally, let us consider an enzyme containing two equivalent active centers interacting with one another. This problem was already examined on p. 112 without using graph theory. Employing the symmetry of the graph and adding the "parallel" branches, we obtain the simplified graph shown in Fig. 46 instead of the diagram in Fig. 37. Here, $F_1 = F_{00}$, $F_2 = F_{10} = F_{01}$, and $F_3 = F_{11}$. We have

$$D_1 = (k_2 + k_{-1}) \, 2 \, (k_4 + k_{-3}), \quad D_2 = 2k_1 S2 \, (k_4 + k_{-3}),$$
$$D_3 = 2k_1 Sk_3 S$$

and

$$v = E \, \frac{k_2 D_2 + 2k_4 D_3}{D_1 + D_2 + D_3} = 2E \, \frac{k_2 K'S + k_4 S^2}{KK' + 2K'S + S^2}, \qquad (I.9)$$

where

$$K = \frac{k_{-1} + k_{-2}}{k_1}, \quad K' = \frac{k_{-3} + k_{-4}}{k_3}.$$

Comparison of this calculation with that given on p. 113 shows that the graph—theory method enables us to avoid direct solution of the equations. This is not so important in the case under consideration, but the advantages of the graph method become greater as the problem becomes more complicated.

The comparison made between the graph method and the method of King and Altman in the article by Vol'kenshtein and Gol'd-shtein [274] showed that the former procedure is far simpler. If we translate the latter method into the language of graph theory, we find that King and Altman evaluated a substantially larger number of trees than is required by the graph method.

Vol'kenshtein and Gol'dshtein [226] applied the graph method to allosteric enzymes. The results obtained were presented in Chapter 12.

The graph method is obviously of general importance for steady-state chemical kinetics. Additional simplifications can be utilized for solving a number of problems by this technique. If an enzyme molecule has several equivalent active centers, the graph of the system is symmetric, as was noted above. If the enzyme has two equivalent centers, the graph consists of two symmetric segments; if there are three equivalent centers, we obtain three equivalent segments, etc. Adding the branches of the graph and taking the symmetry into account, we obtain a single composite vertex instead of a set of equivalent vertices. The concentration of the intermediate compound in the composite vertex equals the enzyme concentrations in each equivalent state multiplied by the number of states. The values of the branches leading to the composite vertex are also multiplied by this number.

In this simplified graph, each vertex is entered by not more than one branch, whose value contains the effector concentration R. The number of vertices z entered by branches containing R is designated as the power of the graph with respect to R (see [281-283]). The values of the trees of the graph contain different powers of R; the greatest power of R cannot be more than z. The numerator and denominator in the expression for the reaction rate can thus be represented as polynomials in R, i.e., in the form

$$v = E \frac{\alpha_0 R^z + \alpha_1 R^{z-1} + \ldots + \alpha_{z-1} R + \alpha_z}{\beta_0 R^z + \beta_1 R^{z-1} + \ldots + \beta_{z-1} R + \beta_z} . \qquad (\text{I.}10)$$

The coefficients in Equation (I.10) are polynomials in the concentrations of other effectors different from R. If R becomes infinitely large, z approximates to the limit α_0/β_0. If R participates in inhibition stages, the denominator is a polynomial in R of higher power than the numerator, and v tends to zero as R increases. The denominator cannot be a polynomial in R of lower power than the numerator.

The power of the polynomial with respect to R can serve to define the number of vertices to which R is attached. It is not necessary to find all the base trees in order to determine the power of the graph with respect to R, since only the highest power of each tree with respect to R need be calculated. One has to plot only those trees in which the effector R appears z times. Such an analysis usually requires only a few minutes.

When the power of D_r has been established, the vertices that can be discarded and those that remain when the enzyme is saturated with this effector become obvious. It is therefore possible to study the mechanism of the reaction from the simplified graph obtained for saturation with one effector.

Let us now consider the application of the graph method to nonsteady-state processes [284]. The derivatives of the concentrations with respect to time do not equal zero and we obtain a system of differential kinetic equations

$$\dot{F}_t(\tau) + F_t(\tau)\sum_{r=1}^{n} a_{tr} = \sum_{s=1}^{n} a_{st}F_s(\tau), \quad t=1, 2, \ldots, n-1. \qquad (I.11)$$

In the general case, where the coefficients a_{tr} and a_{st} depend on the time τ, system (I.11) is not linear and has no precise analytic solution. An approximate solution can be obtained with the aid of analog or digital computers [285]. In some cases, however, the concentrations of the reagents S, I, A, . . . can be treated as constant throughout the course of the reaction. This situation occurs if the concentrations of the reagents substantially exceed the enzyme concentration or if they are artificially held constant.

As an initial condition, it is natural to assume that only the free-enzyme concentration does not equal zero at the starting point, i.e.,

$$F_1(0) = E; \quad F_{t\neq1}(0) = 0. \qquad (I.12)$$

We then apply the Laplace—Carson integral transform [286] to the system of linear equations (I.11). In this case, the functions of time $F(\tau)$ are replaced by their transformants $F^*(q)$, while the derivative with respect to time of $F(\tau)$ are replaced by $qF^*(q) - qF(0)$. The values of q have the meaning of inverse time. The Laplace—Carson transform yields a system of algebraic equations in the transformant $F^*(q)$ instead of the differential equations (I.11):

$$F_t^*(q)\sum_{r=1}^{n} a_{tr}^* = \sum_{s=1}^{n} a_{st}^*F_s^*(q), \quad t=1, 2, \ldots, n-1. \qquad (I.13)$$

Here we have used the initial conditions (I.12) and the condition of constancy of the concentration sum (I.3).

System (I.13) is identical in form to system (I.2). The coefficients a^* are identical to the coefficients a in the equations for steady-state reactions (I.2) for all stages except those leading to the initial state, i.e.,

$$a_{tj}^* = \begin{cases} a_{tj}, & j\neq 1, \\ a_{t1}+q, & j=1. \end{cases} \qquad (I.14)$$

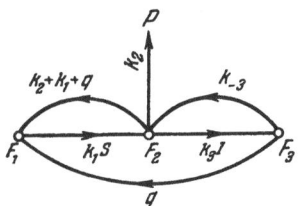

Fig. 47. Graph for nonsteady-state noncompetitive inhibition.

In the language of graph theory, this means that system (I.13) is equivalent to a graph for nonsteady-state stages.

The graph for the nonsteady-state reaction is obtained from that for the steady-state reaction by addition of new branches from each vertex to the initial vertex 1. The value of each new branch is q. The graph so constructed is solved in the usual manner from Equation (I.4). Having solved the graph, we find the Laplace—Carson transformant of the reaction rate v*(q). It has the form of the rational fractional function

$$v^{\bullet}(q) = E\,\frac{\alpha_2 q^{n-2} + \alpha_3 q^{n-3} + \ldots + \alpha_{n-2}}{q^{n-1} + \beta_2 q^{n-2} + \ldots + \beta_{n-1}}\,,\qquad (I.15)$$

where n is the number of vertices in the graph.

Assuming q to be approximately the inverse time, we can compare different processes with respect to the time required for establishment of the steady state. As this state is approached, $q \to 0$ and we obtain the steady-state value of v:

$$v^{\bullet}(0) = v(\infty) = E\,\frac{\alpha_{n-2}}{\beta_{n-2}}\,.\qquad (I.16)$$

The sense α_{n-2} and β_{n-2} thus becomes clear. In order to investigate the initial period of the process, we leave only the highest power of q in the expression for v*(q). For example,

$$v^{\bullet}(q \to \infty) = E\,\frac{\alpha_2}{q}\,.\qquad (I.17)$$

Then,

$$v(\tau \to 0) = E\alpha_2\tau.\qquad (I.18)$$

The value of α_2 is consequently determined from the initial slope of the curves for $v(\tau)$. On the other hand, this quantity can be determined from the graph. Analysis of the initial period of the process is extremely simple. The original of $v(\tau)$ is precisely determined from the transformant v*(q) with the aid of tables of Laplace—Carson transforms, provided the number of vertices in

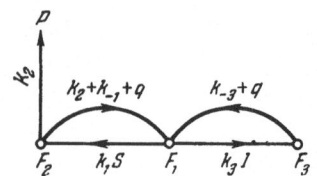

Fig. 48. Graph for nonsteady-state
competitive inhibition.

the graph does not exceed four,
since this number is associated
with determination of the roots of
the denominator in algebraic form.
Numerical determination of the original or special procedures are required for higher-order graphs.

Let us consider the solutions
of the three problems previously worked out for the steady state.
The graph for noncompetitive nonsteady-state inhibition has the
form shown in Fig. 47. It differs from the graph in Fig. 44 in having
two additional branches.

The base determinants are

$$D_1 = (q + k_2 + k_{-1})k_{-3} + k_3Iq + (q + k_2 + k_{-1})q,$$

$$D_2 = k_1Sk_{-3} + qk_1S, \quad D_3 = k_1Sk_3I.$$

The transformant of the reaction rate has the form

$$v^*(q) = \frac{Ek_2D_2}{\sum_r D_r} = k_1k_2ES \frac{q + k_{-3}}{q^2 + 2Aq + B}, \qquad (\text{I.19})$$

where $2A = k_2 + k_{-1} + k_{-3} + k_3I + k_1S$ and $B = (k_2 + k_{-1})k_{-3} + k_1Sk_{-3} + k_1Sk_3I$. Its original, found from the table [286],equals

$$v(\tau) = v_{\text{steady}} - k_1k_2ES\left[\frac{q_1 + k_{-3}}{q_1(q_2 - q_1)}e^{q_1\tau} + \frac{q_2 + k_{-3}}{q_2(q_1 - q_2)}e^{q_2\tau}\right]. \qquad (\text{I.20})$$

Here, $q_{1,2} = -A \pm \sqrt{A^2 - B}$ are the roots of the denominator of
Equation (I.19). Both roots are negative. The function $v(\tau)$ varies
from $v(0) = 0$ to $v(\infty) = v_{\text{steady}}$. The character of the approximation
to the steady-state value depends on the sign of the expression
under the radical, $A^2 - B$. If $A^2 - B \geq 0$, the change is monotonic;
if the constants and concentrations are such that $A^2 - B < 0$, we
obtain damped oscillations. This is due to the fact that an imaginary factor appears in the exponents.

Figure 48 shows the graph for competitive nonsteady-state
inhibition.

The base determinants are

$$D_1 = (k_{-3} + q)(k_{-1} + k_2 + q), \quad D_2 = (k_{-3} + q) k_1 S,$$
$$D_3 = (k_{-1} + k_2 + q) k_3 I.$$

The transformant of the reaction rate has the form

$$v^*(q) = k_1 k_2 ES \frac{q + k_{-3}}{q^2 + 2Aq + B} \cdot \tag{I.21}$$

The values of A are the same for competitive and noncompetitive inhibition, while those of B differ. For competitive inhibition,

$$B = k_{-3}(k_2 + k_{-1}) + k_{-3} k_1 S + (k_{-1} + k_2) k_3 I.$$

A simple calculation shows that $A^2 - B$ cannot be negative for noncompetitive inhibition, which precludes the possibility of oscillations in this case. For competitive inhibition,

$$A^2 - B = \frac{1}{4}(k_{-1} + k_2 - k_{-3} + k_1 S + k_3 I)^2 - k_3 I (k_{-1} + k_2 - k_{-3}). \tag{I.22}$$

If $k_{-1} + k_2 > k_{-3}$, $A^2 - B$ can be less than zero under certain conditions and oscillations with the frequency $\omega = B - A^2$ are possible. In this case, the expression for the reaction rate takes the form

$$v = v_{\text{steady}} - k_1 k_2 SE \frac{e^{-A\tau}}{B}\left[(Ak_{-3} - B)\frac{\sin \omega\tau}{\omega\tau} + k_{-3}\cos \omega\tau\right]. \tag{I.23}$$

Calculations for enzymes of the peroxidase type [287] have shown that the oscillations in such systems are of low amplitude and cannot as yet be detected experimentally.

Graph theory makes it possible to interpret oscillations in enzymatic systems. It is known that a graph represents flows [275]. The flow from vertex i to vertex j along the branch i → j equals the product of the nodal value F_i (the concentration in the vertex) and the probability of the stage a_{ij}. Under presteady-state conditions, intermediate compounds of the enzyme are formed until their concentrations reach the steady-state level; in this case, part of the flow from each vertex returns to the initial vertex. The counterflows cause appearance of new branches with the value q. If the graph has no fewer than three vertices, there are at least two branches with the value q. There can then be a situation in

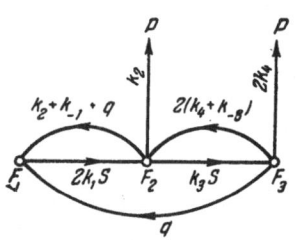

Fig. 49. Graph for nonsteady-state cooperative system with two equivalent active centers.

which loss of a compound in one intermediate state as a result of counterflow leads to an increase in forward flow in another intermediate state.

This might be the key to an understanding of the variations in the rates of enzymatic reactions during the presteady-state period. The rapid damping of the oscillations might be due to the fact that the forward flows exceed the counterflows.

Figure 49 shows a graph for a problem involving two interacting active centers in the nonsteady state. In this case, the base determinants are

$$D_1 = (k_2 + k_{-1} + q)\,2\,(k_4 + k_{-3}) + q\,(k_{-1} + k_2 + q) + qk_3S,$$
$$D_2 = 2k_1S2\,(k_{-3} + k_4) + q2k_1S, \quad D_3 = 2k_1Sk_3S.$$

The transformant of the reaction rate has the form

$$v^{\bullet}(q) = 2k_1k_2ES\,\frac{q+C}{q^2 + 2Aq + B}, \tag{I.24}$$

where

$$2A = 2k_4 + 2k_{-3} + k_{-1} + k_2 + k_3S + 2k_1S,$$

$$B = 2\,(k_{-2} + k_{-1})\,(k_4 + k_{-3}) + 4k_1S\,(k_{-3} + k_4) + 2k_1k_3S^2,$$

$$C = 2\,(k_{-3} + k_4) + 2Sk_4k_3/k_2.$$

This expression thus differs from those for the previous rates in the values of A, B, and C. The originals of the rate, like its transformants for graphs with three vertices, have the same form. The original of the rate for the problem with two centers is

$$v(\tau) = v_{\text{steady}}\ - 2k_1k_2SE\left[\frac{q_1+C}{q_1\,(q_2-q_1)}\,e^{q_1\tau} + \frac{q_2+C}{q_2\,(q_1-q_2)}\,e^{q_2\tau}\right], \tag{I.25}$$

where, as before, $q_{1,2} = -A \pm \sqrt{A^2 - B}.$

For graphs containing a larger number of vertices than that considered above, the numerator and denominator of the rate transformant v*(q) are polynomials in q of higher power. However, since q is a quantity roughly inverse to the time τ, we can limit ourselves to the quadratic polynomial in q in the denominator for determining the initial time. The analysis is made by the method described above. It must be noted that the trend of the process with time determines the values of A and B in the denominator. In order to determine them by the graph method, it is necessary to find only those trees containing the first power of q or not containing q at all. Only the first powers of q in the rate expressions need be taken into account in investigating the initial period of the process. We then obtain identical expressions for competitive and noncompetitive inhibition

$$v^{\bullet}(q) \approx k_1 k_2 ES \frac{1}{q+2A}, \tag{I.26}$$

or

$$v(\tau) \approx k_1 k_2 ES \frac{1}{2A} (1 - e^{-2A\tau}). \tag{I.27}$$

In order to determine the initial slope of the curves for $v(\tau)$, it is sufficient to consider the highest power of q. We obtain

$$v^{\bullet}(q) \approx k_1 k_2 ES/q \quad \text{or} \quad v(\tau) \approx k_1 k_2 ES\tau.$$

The technique in question can thus be employed to determine the kinetic constants of the initial slope of the curves for $v(\tau)$. The graph method permits calculation both of complex enzymatic reactions and of other complex catalytic or chain reactions [280]. The broad class of complex reactions can be represented in the form of a sequence of stages [288]:

$$X_0 + A_0 \rightleftarrows X_1 + B_1,$$
$$X_1 + A_1 \rightleftarrows X_2 + B_2, \ldots, X_s + A_s \rightleftarrows X_0 + B_0.$$

Here, A_0, A_1, . . . designate one or more initial-compound molecules or no such molecules; the same is true of the products B_0, B_1, . . . , while X_0, X_1, . . . designate no more than one intermediate-compound molecule. This system of stages can be used to plot a graph. The absence of intermediate compounds in any stage is represented by a "null compound" with a concentration of one. The vertices of the graph indicate the intermediate or null compounds, while the branches connecting these vertices correspond to the stages.

Temkin investigated the kinetics of complex reactions [289] and subsequently used graphs to determine the "paths" of the independent reactions in a complex reaction system [290]. However, the potentialities of graph theory were not adequately utilized in this case. Graph theory makes it possible to obtain algorithms for solving problems in the kinetics of complex steady-state reactions [280]. Each vertex is comparable to the concentration of an intermediate compound. The concentration of the "null compound" in a vacant vertex is one. Two-directional branches are comparable to reversible stages. The values of the branch in each direction equal the probability of the stage. For example, for the sequence (*), we have $w_{12} = k_{12}A$, $w_{21} = k_{21}B_2$, etc. With this representation, the stage r_{ij} in the direction i → j or the flow in this direction equals the product of the i-th nodal value x_i and the value of the branch w_{ij}, i.e., $r_{ij} = x_i w_{ij}$. We can derive a simple rule for the ratio of two nodal values: $x_i/x_j = D_i/D_j$, where D_i and D_j are the base determinants. Taking the vacant vertex 0 as j, we obtain $x_i = D_i/D_0$ for the intermediate-compound concentrations. It is of interest to determine the reaction rates along independent routes. The routes of a complex reaction correspond to closed circuits in its graph. The mechanism of a complex reaction is usually represented by a planar graph. The shortest closed circuits, i.e., the circuits containing the minimum number of branches, yield independent main routes. The number of shortest closed circuits gives the number of independent routes. The positive direction of the shortest closed circuits, e.g., clockwise, is predetermined. The positive direction of any circuit including line Hamiltonians (passing through each vertex of the graph once) is so defined. It can be demonstrated that the rate r_C over the main route C (the shortest closed circuit C) is calculated from the simple and graphic rule

$$r_C = r_C^+ + r_C^- = \frac{1}{D_0} \sum_{n=0}^{N} (c_n^+ - c_n^-) D_{c_n}. \qquad (\text{I}.28)$$

Here, r_C^+ and r_C^- are the rates over circuit C in the positive and negative directions, respectively, c_0^+ is the value of circuit C in the positive direction, i.e., the product of the values of the circuit

branches, c_n^+ (n = 1, 2, . . . , N) are the values in the positive direction of the circuits surrounding circuit C and having branches
in common with it, down to line Hamiltonians, c_n^- are the same
values in the negative direction, D_{C_n} are the subgraphs remaining
after compression of circuit C to a point (the base), and D_0 is the
base determinant of a graph rooted in a vacant vertex.

ANOMALOUS DISPERSION OF MAGNETIC ROTATION

In 1846, Faraday discovered a phenomenon that he called magnetic rotation of the plane of polarization of light. When light polarized parallel to a magnetic field passes through any substance, the polarization plane of the light wave is rotated by an angle proportional to the thickness of the layer of material and the strength of the magnetic field H

$$\varphi = VlH.$$
(II.1)

Here V is the magnetic-rotation constant or Verdet's constant. Faraday wrote: ". . . I was ultimately able to magnetize and electrify a light beam and illuminate a magnetic line of force" [291]. This is naturally a figurative statement. In actuality, the Faraday effect essentially consists in the action of a magnetic field on a substance, so that it acquires circular birefringence (see p. 59) and circular dichroism (see p. 68).

The Faraday effect differs radically from natural optical activity. The latter results from asymmetry inherent in the medium itself (molecular or crystalline); in magnetic rotation, on the other hand, asymmetry is induced by the magnetic field, whose strength is an axial vector. The Faraday effect is therefore observed in all compounds, regardless of the symmetry or asymmetry of their molecules. It is obvious that, having different causes, the Faraday effect can yield different information on the structure of a compound than study of optical activity. In seeking new ways to investigate the structure of biopolymers, it is natural to turn to the Faraday effect.

The optical properties of a molecule are determined by its polarizability [107, 108]. The Faraday effect is obviously a manifestation of the influence of a magnetic field on polarizability. Having acknowledged the existence of this influence, let us expand the components of the polarizability tensor into a series in the magnetic–field strength

$$a_{\sigma\tau} = \alpha_{\sigma\tau} + \sum_{\rho} \alpha_{\sigma\tau,\rho} H_\rho + \frac{1}{2} \sum_{\rho,\nu} \alpha_{\sigma\tau,\rho\nu} H_\rho H_\nu + \ldots \qquad (\text{II.2})$$

Here, σ, τ, ρ, and ν are the coordinates of the system embodied in the molecule (ξ, η, ζ); H_ρ and H_ν are the components of the magnetic–field vector and $\mathbf{H}$ is the axial vector, which is equal to the curl of the potential vector

$$\mathbf{H} = \mathrm{curl}\,\mathbf{A}. \qquad (\text{II.3})$$

The quantity $\alpha_{\sigma\tau,\rho}$ is consequently asymmetric with respect to permutation of the subscripts σ and τ. Actually,

$$\alpha_{\xi\eta,\zeta} = \left(\frac{\partial a_{\xi\eta}}{\partial H_\zeta}\right)_{H=0} = \left(\frac{\partial a_{\xi\eta}}{\partial\left(\dfrac{\partial A_\eta}{\partial\xi} - \dfrac{\partial A_\xi}{\partial\eta}\right)}\right)_{H=0} = -\alpha_{\eta\xi,\zeta}.$$

Since it follows from the law of conservation of energy that tensor $\alpha_{\sigma\tau}$ is a Hermitian, $\alpha_{\sigma\tau,\rho}$ should be imaginary [107]. The effect is linear with respect to H and we can consequently limit ourselves to two terms on the right side of Equation (II.2). Let us assume that the field $\mathbf{H}$ is directed along the z axis of a spatially stationary coordinate system. The relationship between the strength and the electrical induction in a substance placed in the field has the form

$$\left.\begin{aligned} D_x &= \varepsilon E_x - i\varepsilon' E_y, \\ D_y &= i\varepsilon' E_x + \varepsilon E_y, \\ D_z &= \varepsilon E_z, \end{aligned}\right\} \qquad (\text{II.4})$$

where (for a gas),

$$\varepsilon' = i\varepsilon_{xy} = 4\pi i N_1 a_{xy}. \qquad (\text{II.5})$$

The quantity ε' arises as a result of the presence of imaginary asymmetric polarizability components; a_{xy} is imaginary and ε' is real.

If a planar light wave propagates along the z axis, it is described by the equations

$$\left.\begin{array}{l} D_x = n^2 E_x, \\ D_y = n^2 E_y, \\ D_z = 0. \end{array}\right\} \tag{II.6}$$

Solving Equations (II.4) and (II.6) simultaneously, we find

$$\left.\begin{array}{l} (\varepsilon - n^2) E_x - i\varepsilon' E_y = 0, \\ i\varepsilon' E_x + (\varepsilon - n^2) E_y = 0, \end{array}\right\} \tag{II.7}$$

whence we obtain two values for the refractive index of the medium:

$$n^2 = \varepsilon \pm \varepsilon'. \tag{II.8}$$

The solution with a plus sign corresponds to $E_y/E_x = i$, i.e., light circularly polarized to the left, while the solution with a minus sign corresponds to $E_y/E_x = -i$, i.e., light circularly polarized to the right. A medium with such properties exhibits circular birefringence characterized by the difference in refractive indices

$$\Delta n = n_r - n_l \approx \frac{\varepsilon'}{n}. \tag{II.9}$$

The corresponding rotation of the polarization plane is [compare Equation (19) on p. 59]

$$\varphi = \frac{\pi}{\lambda} \Delta n l = \frac{\pi \varepsilon'}{\lambda n} l. \tag{II.10}$$

Comparing Equations (II.10) and (II.1), we find

$$V = \frac{\pi \varepsilon'}{\lambda n H}, \tag{II.11}$$

where ε' is proportional to H.

This is the phenomenological theory. The polarizability is in turn expressed by the frequency and probability of spectral transitions between electron levels, which are determined by p_{0j} (see p. 60). The influence of a magnetic field on polarizability thus reduces to its influence on the energy levels and transition probabilities. A magnetic field causes splitting of energy levels (the Zeeman effect). The simplest classical model of the electron

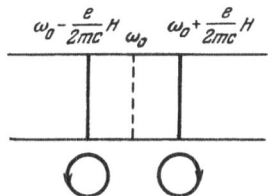

Fig. 50. Simple longitudinal
Zeeman effect.

treats it as a harmonic oscillator. If it oscillates with an angular frequency ω_0 in the absence of a magnetic field, a longitudinal Zeeman effect occurs in the presence of a field directed parallel to the light beam; the spectral line with the frequency ω_0 is split into two lines circularly polarized to the right and left (Fig. 50). The magnitude of the splitting is $2|\omega_L|$, where ω_L is the Larmor-precession frequency

$$\omega_L = -\frac{eH}{2mc}. \tag{II.12}$$

The actual frequency for left-polarized light is thus $\omega_0 - \omega_L$ rather than ω_0, while that for right-polarized light is $\omega_0 + \omega_L$ rather than ω_0. The circular birefringence can be represented by the quantity

$$\Delta n = \left(\frac{\partial n}{\partial \omega}\right)_{\omega_L=0} 2\omega_L = -\frac{eH}{mc}\left(\frac{\partial n}{\partial \omega}\right)_{\omega_L=0}, \tag{II.13}$$

or, converting to wavelengths $\lambda = 2\pi c/\omega$, we obtain

$$\Delta n = \lambda^2 \left(\frac{\partial n}{\partial \lambda}\right)_{H=0} \frac{eH}{2\pi mc^2}; \tag{II.13a}$$

substituting into Equation (II.1), we find

$$\varphi = \frac{eH}{2mc^2} \lambda \left(\frac{\partial n}{\partial \lambda}\right)_{H=0} l, \tag{II.14}$$

whence

$$V = \frac{e\lambda}{2mc^2}\left(\frac{\partial n}{\partial \lambda}\right)_{H=0}. \tag{II.15}$$

This is the Becquerel equation. The Verdet constant is expressed by the derivative of the refractive index with respect to wavelength, i.e., the dispersion function of the Faraday effect is the derivative with respect to wavelength of the dispersion function of ordinary refraction. This result is also yielded by the classical theory, provided that the compound is diamagnetic and that only the normal Zeeman effect is observed in its spectrum. The few measurements that have been made of the dispersion of magnetic rotation (DMR) outside the absorption region for a number of compounds are in rough agreement with the Becquerel equation.

The classical problem of the polarizability of an electron (harmonic oscillator) in a magnetic field and thus exposed to the Lorentz force and the light-wave field reduces to solution of the electron-movement equation

$$m\ddot{r} + kr = \frac{e}{c}[\dot{r}H] + eE_0 e^{i\omega t}. \tag{II.16}$$

Strict solution of this problem (neglecting damping, i.e., light absorption) yields [107]

$$\varphi = \frac{\omega^2 l \omega_L}{cn} 4\pi \left(\frac{n^2+2}{3\cdot}\right)^2 N_1 \frac{e^2}{m} \sum_j \frac{f_j}{(\omega_j^2 - \omega^2)^2 - 4\omega_L^2 \omega^2}, \tag{II.17}$$

for a system of such oscillators and, since ω differs materially from ω_j, $\omega_j^2 - \omega^2 \gg 2\omega_L\omega$ (ω_L is small until H reaches tens of thousands of gausses). We can consequently ignore the term $4\omega_L^2 \omega^2$ in the denominator of the dispersion equation. Magnetic rotation has the specific, symmetric dispersion function

$$\varphi \sim \sum_j \frac{f_j}{(\omega_j^2 - \omega^2)^2}, \tag{II.17a}$$

while ordinary polarizability is expressed by the dispersion sum

$$\sum_j \frac{f_j}{\omega_j^2 - \omega^2}.$$

In other words, φ is expressed by the derivative of n with respect to frequency (or wavelength).

The origin of symmetric DMR is readily seen if we consider the natural-absorption region. The curves for the dispersion n, which correspond to the components of the Zeeman splitting, are identical but displaced relative to one another by $2\omega_L$ (Fig. 51). The value of φ is expressed by the difference in these curves. The difference Δn has the form of a single curve with a minimum (Fig. 51a) or with two minima at large H, i.e., ω_L (Fig. 51b). Since the natural widths of the molecular bands are far greater than ω_L, only the first case can theoretically occur in the molecular spectra.

This type of anomalous dispersion of magnetic rotation (ADMR) was first observed by Macaluso and Corbino in sodium vapor in 1898 ([292]; see also [107, 293]). Symmetric ADMR is

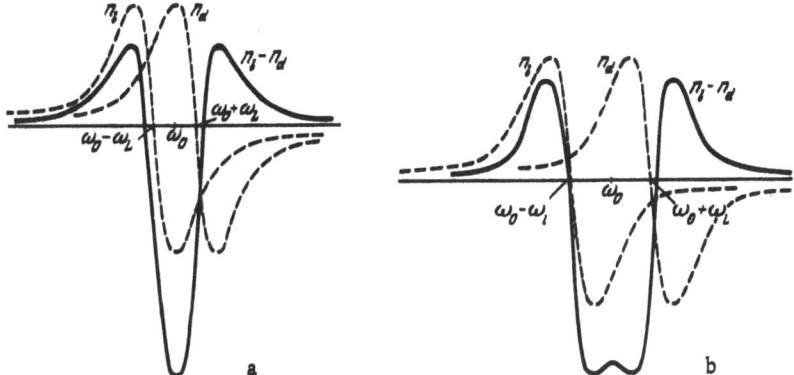

Fig. 51. Anomalous dispersion of magnetic rotation: Macaluso—Corbino phenomenon.
a) Weak magnetic field; b) strong field.

called the Macaluso—Corbino phenomenon. It should be quite pronounced and very sensitive to the positions and intensities of the spectral lines; in essence, there is an interference, or difference effect within a given absorption line (band). The classical treatment of magnetic rotation is based on consideration of the inverse Zeeman effect, i.e., the Zeeman effect in the absorption spectrum. Direct observation of the Zeeman effect in molecular spectra is greatly hampered by the small magnitude of the splitting. On the other hand, the Faraday effect is readily observed, since we are not dealing with ADMR.

All the foregoing pertains to diamagnetic compounds. Paramagnetic compounds exhibit a Faraday effect governed by the orientation of the molecular magnetic moments μ and hence are temperature-dependent. In this case, the classical electronic theory leads to the formula [107]

$$\varphi \approx \frac{\omega l \pi}{cn} \frac{2\mu H}{3kT} \left(\frac{n^2 + 2}{3} \right)^2 N_1 \frac{e^2}{m} \sum_j \frac{f_j}{\omega_j^2 - \omega^2} . \tag{II.18}$$

The dispersion function is asymmetric and the Becquerel equation consequently cannot be valid for paramagnetic compounds.

The rigorous theory of magnetic rotation is naturally a quantum-mechanical theory. It was developed on the basis of perturbation theory and has been considered in a number of articles ([294-296]; see also [107, 297]).

While there is far-reaching correspondence between the classical and quantum theories of molecular electrical properties (polarizability and dipole moment), no such correspondence exists in the treatments of magnetic phenomena. The classical theory of diamagnetism deals solely with the diamagnetism resulting from Larmor precession. Quantum mechanics yields a similar physical result only for spherically symmetric systems (atoms). In the rigorous theory, i.e., quantum mechanics, the dielectric permeability of molecules is represented by the sum of the precessive diamagnetism and the so-called van Vleck paramagnetism, or temperature-independent paramagnetism, i.e.,

$$\chi = \chi_1 + \chi_2 = -N_A \frac{e^2}{6mc^2} \sum_n (r_n^2)_{00} + N_A \frac{2}{3} \sum_j{}' \frac{\mu_{0j}^2}{h\nu_j} \qquad (\text{II}.19)$$

(see, for example, [298]). Here n is the electron index of the molecule, r is the electron coordinate, μ_{0j} is the matrix element of the magnetic moment for the transition with frequency ν_j between the ground level 0 and the excited level j. The first term in Equation (II.19) represents the precessive diamagnetism and is analogous to the classical expression. The second term, which is positive, describes the quantum-mechanical effect of the electron-shell deformation caused by the external magnetic field and, in this sense, is similar to polarizability.

The lack of analogy between the classical and quantum theories of magnetic properties is also demonstrated by the fact that the Zeeman effect is associated with Larmor precession and hence with diamagnetism in the classical theory and with the presence of natural orbital and spin moments, i.e., with paramagnetism, in the quantum theory. This difference results from the nonrigorous and inconsistent character of the classical theory of diamagnetism, which tacitly assumes the electron orbitals in the atoms to have a quantum character. The strict classical theory should lead to zero diamagnetic permeability [299].

Equation (II.19) and the quantum-mechanical expression for the Verdet constant accordingly consists of two terms

$$V = V_1 + V_2, \qquad (\text{II.20})$$

where

$$V_1 = \frac{16\pi^2 N_1 v^2}{3h^2} \sum_{(\xi,\,\eta,\,\zeta)} \sum_j \frac{v_j \left[(\mu_\zeta)_{jj} - (\mu_\zeta)_{00} \right] \operatorname{Im} \left[(p_\xi)_{0j} (p_\eta)_{j0} \right]}{(v_j^2 - v^2)^2}, \qquad (\text{II.21})$$

$$V_2 = \frac{8\pi^2 N_1 v^2}{3h^2} \sum_{(\xi,\,\eta,\,\zeta)} \sum_j \sum_k {}' \left\{ \frac{\operatorname{Im} \left[(\mu_\zeta)_{jk} (p_\xi)_{0j} (p_\eta)_{k0} + (\mu_\zeta)_{kj} (p_\xi)_{0k} (p_\eta)_{j0} \right]}{v_{kj} (v_j^2 - v^2)} \right.$$

$$\left. + \frac{\operatorname{Im} \left[(\mu_\zeta)_{k0} (p_\xi)_{0j} (p_\eta)_{jk} + (\mu_\zeta)_{0k} (p_\xi)_{kj} (p_\eta)_{j0} \right]}{v_k (v_j^2 - v^2)} \right\}. \qquad (\text{II.22})$$

Summation is carried out for the cyclic permutations of the coordinates ξ, η, and ζ in the system embodied in the molecule.

A third term is added to the Verdet constant (II.20) for paramagnetic compounds:

$$V_3 = \frac{8\pi^2 N_1 v^2}{3hkT} \sum_{(\xi,\,\eta,\,\zeta)} \sum_j \frac{(\mu_\zeta)_{jj} \operatorname{Im} \left[(p_\xi)_{0j} (p_\eta)_{j0} \right]}{v_j^2 - v^2}. \qquad (\text{II.23})$$

The quantity V_1 corresponds to the precessive diamagnetism χ_1, while the quantity V_2 corresponds to the temperature-independent paramagnetism χ_2. The quantity $(\mu)_{00}$ in Equation (II.21) is zero for diamagnetic compounds.

The Becquerel equation is obviously rigorous only for diamagnetic compounds for which $V_2 \ll V_1$. No appropriate investigations have been conducted: indeed, the problem is here formulated for the first time.

It is also apparent that the Macaluso—Corbino phenomenon corresponds to the term V_1 and not V_2. In other words, this phenomenon is possible only for those absorption bands governed by transitions to an energy level where the magnetic moment μ_{jj} differs from zero, i.e., a paramagnetic level.

The term V_2 is substantially more complex than V_1, since it is governed by electron transitions involving three rather than two levels.

This is a brief outline of the theory of the molecular Faraday effect. Its current state is comparable to that of the theory of optical activity after derivation of the Rosenfeld equation (see p. 61) but before approximation methods for calculating molecular rotary power had been developed. There are as yet no procedures for calculating magnetic rotation.

Until recently, very little attention was paid to the Faraday effect in molecular systems (in contrast to that in certain crystalline compounds). This may have been psychologically associated with the work of Allison, which was conducted during the nineteen-twenties but proved to be in error ([107], p. 602). When it was found that the Faraday effect did not provide the unique opportunities for studying the structure of matter ascribed to it by Allison, interest in magnetic rotation dropped sharply.

On the other hand, it is obvious that study of DMR and especially ADMR can theoretically provide new and valuable information on molecular structure, particularly on macromolecular conformations. The numerators of Equations (II.21) and (II.22) contain different derivatives of three vectors: one magnetic and two electric. Since the directions of these vectors are fixed in a molecule of given structure, the DMR should be very sensitive to changes in molecular conformation. This was noted by Tinoco and his colleagues, who made rough calculations for biopolymers [140].

The ADMR should be even more sensitive. This phenomenon can be referred to as the spectrum of magnetic rotation; however, strictly speaking, magnetic-rotation spectra are investigated in a different manner. Instead of determining the angle of rotation of the polarization plane, one determines the intensity of the light of a given wavelength transmitted through a crossed polarizer and analyzer, between which the compound in question is held in a longitudinal magnetic field. This method has yielded interesting results for gases [300-302]; a theory of magnetic-rotation spectra has been worked out by Hameka [303].

At the time of writing, research on ADMR has been conducted at only a few facilities, including Shashoua's laboratory in the USA and the author's laboratory. Shashoua described an experimental procedure and measured the ADMR in a number of low-molecular weight compounds [304]. His later articles were devoted to ADMR in porphyrin compounds and use of this technique to investigate cytochrome oxidation [305]. He demonstrated the high sensitivity of the method and specifically established that it is possible to make quantitative studies of the kinetics of cyto-

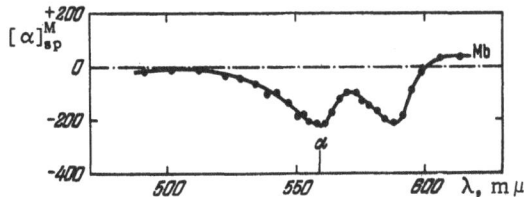

Fig. 52. Anomalous dispersion of magnetic rotation in
α-band of myoglobin. Here and in the following figures,
the vertical lines on the abscissa indicate the positions of
the absorption maxima.

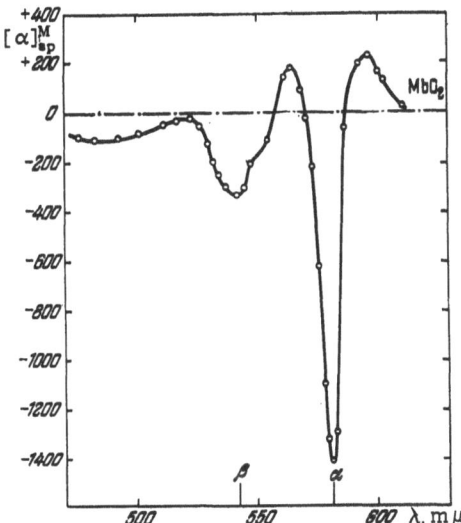

Fig. 53. ADMR in α-band of MbO₂.

chrome oxidation. However, he did not give any theoretical interpretation of his results.

The present author and his colleagues investigated the oxygenation of myoglobin and hemoglobin [249, 306].

Figure 52 shows a curve representing the ADMR in the α-
band of myoglobin. Here and henceforth, the specific magnetic rotation $[\alpha]_{sp}^{M}$ is for a field strength of 10,000 G. Figure 53 presents
similar data for MbO₂, while Figs. 54 and 55 show the ADMR in
the Soret band for Mb and MbO₂, respectively.

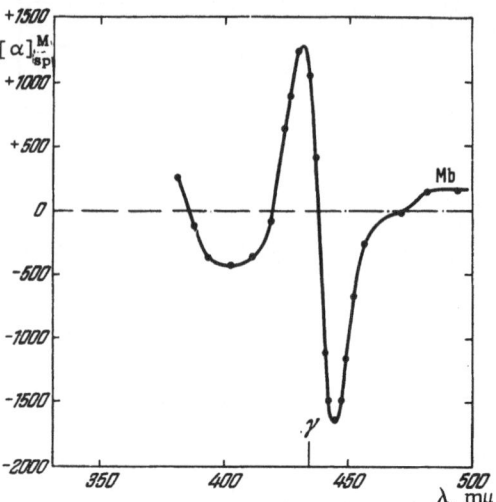

Fig. 54. ADMR in Soret band of Mb.

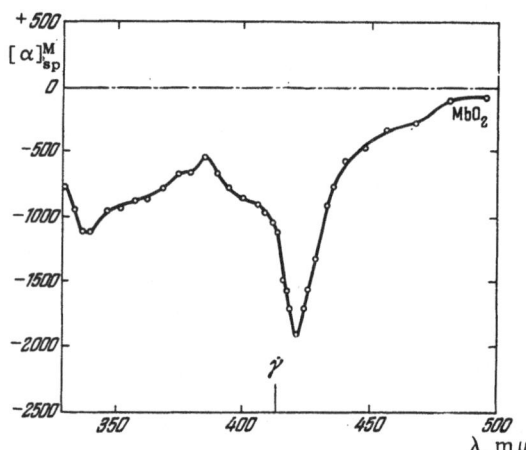

Fig. 55. ADMR in Soret band of MbO_2.

First of all, we must note the differing character of the ADMR in different absorption bands and the severe change it undergoes during oxygenation of myoglobin. Mb is paramagnetic, while MbO_2 is diamagnetic. A pronounced Macaluso—Corbino effect is observed in the α–band of MbO_2; this effect is apparently

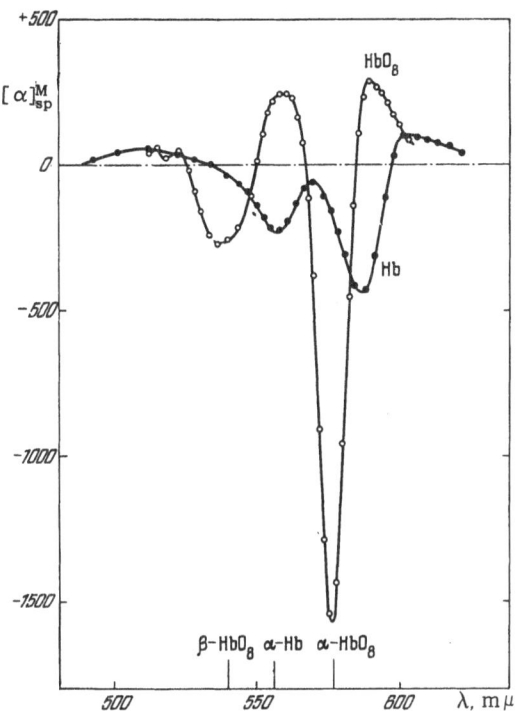

Fig. 56. ADMR in α-bands of Hb and HbO₈.

represented in Fig. 55, in the Soret band of MbO₂. On the other
hand, asymmetric ADMR is observed in the Soret band of Mb; a
far weaker double Macaluso—Corbino effect occurs in the α-band
of Mb.

Figures 56-58 present similar data for hemoglobin and oxygenated hemoglobin. The general character of the curves is the
same as that for myoglobin, but comparison of the ADMR of Mb
and Hb and of MbO₂ and HbO₈ indicates that there are substantial
quantitative differences, which undoubtedly reflect the interaction
between the myoglobin-like subunits in the quaternary structure
of hemoglobin. Thus, for example, the two minima in the ADMR
curve in the vicinity of the Mb α-band (see Fig. 52) are almost
identical, while the corresponding minima in the Hb curve (Fig.
56) differ by a factor of almost two. The ADMR is extremely
sensitive to oxidation of heme-containing proteins and can be employed for quantitative study of this process.

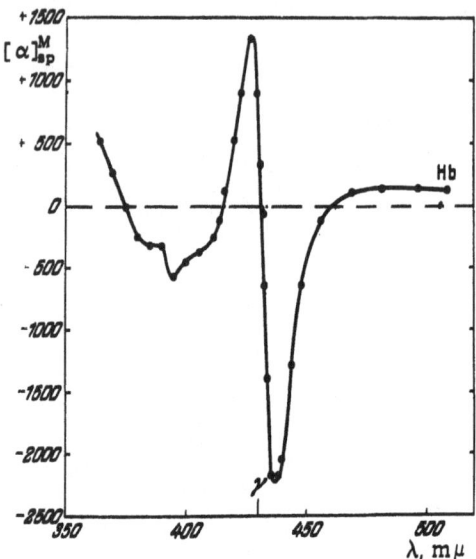

Fig. 57. ADMR in Soret band of Hb.

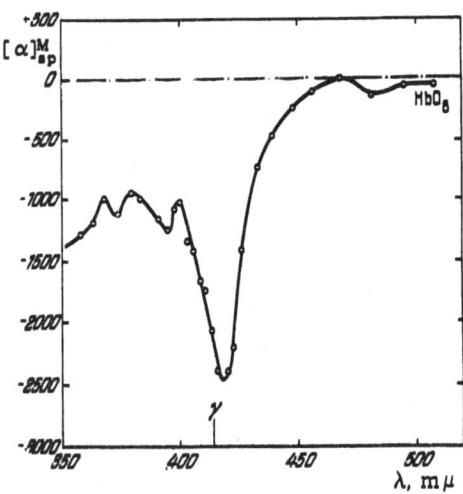

Fig. 58. ADMR in Soret band of HbO₈.

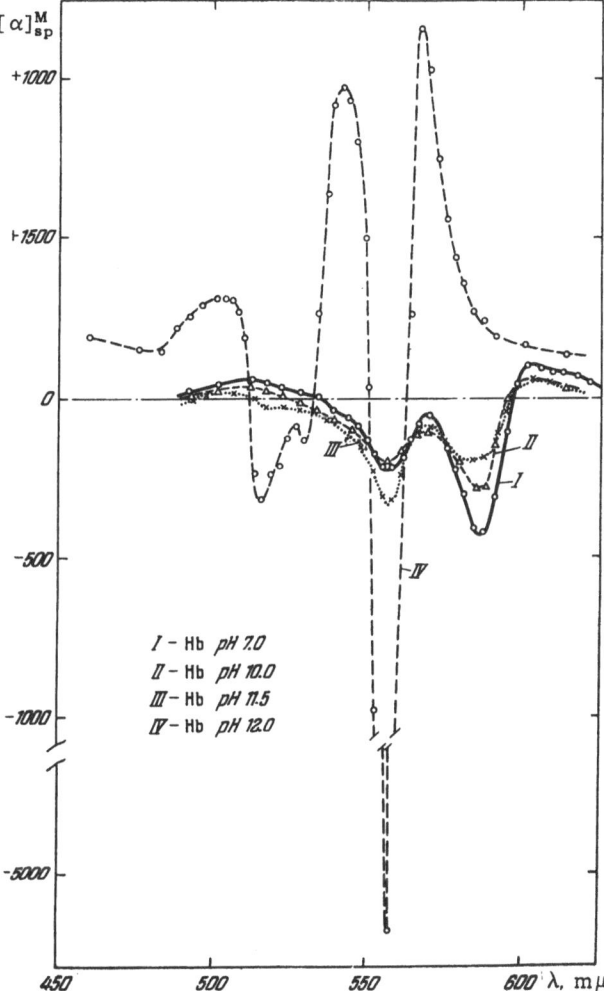

Fig. 59. ADMR in α-band of Hb at different pH's.

A change in medium pH also has a severe effect on the form
of the ADMR curve. Figure 59 shows the trend of the ADMR for
the Hb α-band at four pH's. As can be seen, the ADMR actually
corresponds to the Macaluso—Corbino effect, the short-wavelength
component becoming substantially more intense and the long-wavelength component being attenuated as the pH is raised. At pH 12,
hemoglobin is denatured and the heme passes into the hemochro-
mogen state, which has a very strong effect on the trend of the ADMR
curve.

The results presented above testify to the valuable opportunities offered by the Faraday effect for research on proteins. However, detailed interpretation of ADMR curves will be possible only after a reliable theory has been worked out. The curves obtained by Shashoua [304, 305] and by the present author [306] were to some extent unexpected. Actually, according to Equation (II.21), the Macaluso—Corbino effect can be observed in a diamagnetic compound only if the magnetic moment of the molecule in the excited state j (the transition to this state corresponding to the band ν_j) differs from zero. The magnetic moment can be of either the spin or orbital type.

The orbital moment of a multiatomic molecule is not a constant of motion. If the molecule is linear, the projection of the moment on an axis of infinitely great order C_∞ serves as such a constant. This does not change the situation, since Equation (II.21) contains projections of the moment on the same bases. However, the molecules studied are not linear. If a molecule has an axis of symmetry of finite order, we must speak of the projection of a quasimoment rather than a moment on it. With an even number of electrons (a diamagnetic molecule), the quasimoment projection equals zero for the axes C_1 and C_2 but, beginning with the axis C_3, quasimoment projections with values different from zero appear [307]. No research has been done on the appearance of the quasimoment in the Faraday effect. Moreover, by virtue of the Jahn—Teller theorem, symmetric electron configurations of nonlinear molecules with no less than third-order axes of symmetry are unstable [308].

A porphyrin ring not containing a metal atom at its center has the symmetry D_{2h}, i.e., the axes are only second-order. The symmetry of metalloporphyrins (specifically heme) is D_{4h}. The Macaluso—Corbino effect is observed in both cases, but is more pronounced for metalloporphyrins. The same is true of phthalocyanin and its complexes with metals. Data on heme-containing proteins [305, 306] show that the Macaluso—Corbino effect is stronger in paramagnetic compounds. This effect may be described by Equation (II.21) with $\mu_{jj} = 0$ but $\mu_{00} \neq 0$. However, Shashoua established that paramagnetism or diamagnetism of the ground state in phthalocyanins has no direct influence on the form of the ADMR curve [305].

The Macaluso—Corbino effect in porphyrin compounds has two possible causes. It may be that the paramagnetism of the excited state is of the spin rather than orbital type. In other words, we are dealing with a singlet—triplet transition in the diamagnetic case. This is apparently contradicted by the low intensity of the corresponding transitions in light absorption. Singlet—triplet transitions (intercombinations) are forbidden in electron spectra by the strict selection rules and are allowed only as a result of perturbing spin-orbital interactions. Theoretical evaluations of the intensities of singlet—triplet transitions have been made for only a few cases (see, for example, [309]); the estimates are actually very small. On the other hand, the Macaluso—Corbino effect is observed in the strong absorption bands of porphyrin compounds. This contradiction appears irreconcilable at first glance.

However, the theory of the Faraday effect presented above shows that magnetic rotation bears no direct relationship to the intensity of light absorption: weak absorption bands can be characterized by strong magnetic rotation (see [310]) and particularly by a strong Macaluso—Corbino effect. It is possible that the Macaluso—Corbino effect in porphyrin compounds does result from singlet—triplet transitions too weak to be observed in the absorption spectrum. These transitions occur in the same region as the $\pi\pi^*$-transitions of porphyrin rings; their appearance is completely masked by these strong absorption bands and they are detectable only from the ADMR or magnetic-rotation spectra.

The second alternative is simpler. The effect results from the orbital quasimoment with respect to the axis C_4; the Jahn—Teller splitting of the bands is too small to have any material effect on the observed pattern. The theoretical study of ADMR in porphyrin compounds made by Stephens, Suetaak, and Schatz [311] supports this second hypothesis. Proceeding from the quantum-mechanical theory of the effect, the authors analyzed their experimental data and conclusively attributed them to orbital paramagnetism. The Jahn—Teller effect was not discussed, but the fact that good agreement was achieved between the experimental ADMR curves and those calculated theoretically with appropriate constants indicates that this effect plays no material role. The results obtained agree well with the interpretations of porphyrin electron spectra in the literature. The magnetic-rotation constants for all the absorption bands of coproporphyrin II, zinc hematoporphyrin, phthalocyanin, and magnesium phthalocyanin were determined. In conformity with the theory, the Macaluso—Corbino effect was large in metallic complexes with the symmetry D_{4h}. However, it was also observed in coproporphyrin II and phthalocyanin, compounds that contain no metal atoms and accordingly have the symmetry D_{2h}. This is due to the higher approximate symmetry D_{4h} in the π-electron shell of the molecule. Pershan et al. [312] used similar bases for developing a theory of the ADMR in flat aromatic molecules.

Stephens et al. [311] considered the metal atom only as a factor bringing in symmetry. Nevertheless, it plays a material role in the observed effect. It was recently established [313] that the Macaluso—Corbino effect in the α-absorption band of ferro-hemoglobin differs for the ligands H_2O, NO, CO, and O_2: the ADMR curves are displaced relative to one another and the amplitudes of the minima differ greatly. These results are in good agreement with those obtained in investigating the Mössbauer effect, which provide information on the electronic state of the central Fe atom of heme. The ligand has a direct influence on this state. On the other hand, it has been found that there is a correlation between the amplitudes of the minima in the ADMR curves and the conformational-stability and cooperativity characteristics of proteins (see p. 138). Construction of a theory of ADMR that takes into account the electronic state of the metal atom is a very urgent task. The current theory of ligand fields should be employed for this purpose (see [297]).

The lack of basic differences in the ADMR for paramagnetic and diamagnetic porphyrin compounds can be attributed to the fact that the Macaluso—Corbino effect is substantially stronger than the corresponding paramagnetic effect, i.e., V_2, defined by Equation (II.22), is far greater than V_3, defined by Equation (II.23), near an absorption band. This is understandable, since V_2 is a resonant function of $(\nu_j^2 - \nu^2)^2$ and V_3 is a resonance function of the weaker quantity $\nu_j^2 - \nu^2$.

All the foregoing shows that research on ADMR is of great interest both for molecular spectroscopy and for the study of molecular structure. It can be assumed that substantial ADMR effects should be observed in all cases. There is one possible experimentally unfavorable situation: a strong absorption band and a

small Faraday effect. Nevertheless, study of the ADMR and corresponding dichroism is very promising. The same can be said of
the normal DMR in a spectral region remote from the natural absorption bands.

In this connection, we might point out the valuable opportunities for study of protein transformations afforded by ADMR. As
has already been mentioned, Shashoua used this technique to study
the kinetics of the oxidation of cytochrome c. The results presented above for investigations of hemoglobin and myoglobin open
up the possibility that their oxygenation kinetics can be investigated. Denaturation of these proteins is also clearly manifested
in their ADMR. Comparison of HbO_8 denaturation kinetics observed with the aid of the AORD in the peptide-bond absorption
band in the vicinity of 200 mμ with the denaturation kinetics observed with the aid of the ADMR in the heme absorption band shows
that the rate constant is somewhat different in the latter case. It
has thus been established that unfolding of the helical segments adjoining the heme group occurs at a different rate than unfolding of
the entire protein macromolecule. The ADMR method therefore
permits a differential approach to study of denaturation (see [322]).

A good survey of magnetic rotation was recently published
by Buckingham and Stephens [323].

Application of the DMR and ADMR techniques to study of biopolymers, particularly enzymes, will clarify many problems. The
principal tasks at hand include study of conformational transformations and of the interactions of enzymes with coenzymes, substrates, inhibitors, and other effectors.

REFERENCES

1. G. Lichtenberg, Aphorisms, Nauka, 1964, p. 56.
2. M. V. Vol'kenshtein, Configurational Statistics of Polymeric Chains, Interscience (Wiley), New York, 1963.
3. M. V. Vol'kenshtein, Molecules and Life, Plenum Press, New York, 1969.
4. T. M. Birshtein and O. B. Ptitsyn, Conformations of Macromolecules, Interscience (Wiley), New York, 1966.
5. L. Pauling and R. Corey, Proc. Roy. Soc., B141: 10 (1953).
6. F. Haurowitz, Chemistry and Biology of Proteins, Academic Press, New York, 1950.
7. P. Doty, in: "Biophysical Science," Rev. Mod. Physics, 31:61 (1959).
8. P. Doty and W. Gratzer, in: Polyamino Acids, Polypeptides, and Proteins, edited by M. Stahmann, University of Wisconsin Press, 1962, p. 111.
9. R. Dickerson, The Proteins, Vol. 2, 2nd edition, Academic Press, New York, 1964, p. 603.
10. M. Perutz, Sci. Am., 211(5): 64 (1964).
11. M. Perutz, J. Mol. Biol., 13: 646 (1965).
12. C. Blake, D. Koenig, C. Mair, A. North, D. Phillips, and V. Sarma, Nature, 206: 757 (1965).
13. K. Linderstrøm-Lang, in: Proteins and Enzymes, Stanford University Press, 1952, p. 115.
14. J. Schellmann, C. R. Lab. Carlsberg, 30: 450 (1958).
15. T. Takagi and T. Isemura, J. Biochem. (Japan), 52: 314 (1962).
16. C. Anfinsen, Proc. Natl. Acad. Sci., U.S., 47: 1230 (1961).
17. C. Anfinsen and E. Haber, J. Biol. Chem., 236: 1361 (1961).
18. C. Epstein, R. Goldberger, and C. Anfinsen, Cold Spring Harbor Symp. Quant. Biol., 28: 439 (1963).
19. C. Levinthal, Proc. Natl. Acad. Sci. U.S., 48: 1230 (1962).
20. A. Guzzo, Biophys. J., 5: 809 (1965).
21. E. Blout, in: Polyamino Acids, Polypeptides, and Proteins, edited by M. Stahmann, University of Wisconsin Press, 1962, p. 275.
22. D. Davis, J. Mol. Biol., 9: 605 (1964).
23. B. Havsteen, J. Theoret. Biol., 10: 1 (1966).
24. H. Saroff, J. Theoret. Biol., 9: 229 (1965).

25. J. Klotz, Brookhaven Symp. Theoret. Biol., 13:25 (1960).

26. J. Klotz and J. Franzen, J. Am. Chem. Soc., 84:3461 (1962).

27. O. B. Ptitsyn and A. M. Skvortsov, Biofiz., 10:909 (1965).

28. S. E. Bresler and D. L. Talmud, Dokl. Akad. Nauk SSSR, 43:326, 367 (1944).

29. S. E. Bresler, Biokhimiya, 14:180 (1949).

30. P. A. Rebinder, Surface-Active Agents, Izd. Akad. Nauk SSSR (1961).

31. W. Kauzmann, Advan. Protein Chem., 14:1 (1959).

32. G. Nemethy and H. Scheraga, J. Chem. Phys., 36:3382 (1962); 41:680 (1964);
     J. Phys. Chem., 66:1773 (1962); 67:2888 (1963).

33. D. Poland and H. Scheraga, Biopolymers, 3:275, 283, 305, 315, 335, 357, 369,
     379, 401 (1965).

34. E. V. Anufrieva, T. M. Birshtein, T. N. Nekrasova, O. B. Ptitsyn, and T. V.
     Sheveleva, J. Polymer Sci. (in press).

35. H. Fisher, Proc. Natl. Acad. Sci. U.S., 51:1285 (1964).

36. H. Fisher, Biochim. Biophys. Acta, 109:544 (1965).

37. A. Szent-Györgyi, Bioenergetics, Academic Press, New York, 1957.

38. M. Perutz, J. Kendrew, and H. Watson, J. Mol. Biol., 13:669 (1965).

39. S. J. Singer, Advan. Protein Chem., 17:1 (1962).

40. E. Margoliash, Can. J. Biochem., 42:745 (1964).

41. M. V. Vol'kenshtein, Genetika, 2:54 (1965).

42. G. Braunitzer, K. Hilse, V. Rudloff, and N. Hilschmann, Advan. Protein Chem.,
     19:1 (1964).

43. V. Ingram, The Hemoglobins in Genetics and Evolution, Columbia University
     Press, New York, 1963.

44. Yu. N. Zograf and M. D. Frank-Kamenetskii, in: Biosynthesis of Proteins and
     Nucleic Acids, Nauka, 1965, Chap. 1.

45. M. V. Vol'kenshtein, Nature, 207:294 (1965).

46. J. Trupin, F. Rottman, R. Brimacombe, P. Leder, M. Bernfield, and M. Nirenberg,
     Proc. Natl. Acad. Sci. U. S., 53:807 (1965).

47. M. Nirenberg, P. Leder, M. Bernfield, R. Brimacombe, J. Trupin, F. Rottman,
     and C. O. Neal, Proc. Natl. Acad. Sci. U. S., 53:1161 (1965).

48. R. Brimacombe, J. Trupin, M. Nirenberg, P. Leder, M. Bernfield, and T. Jaonni,
     Proc. Natl. Acad. Sci. U. S., 54:954 (1965).

49. M. V. Vol'kenshtein, Genetika, No. 4:119 (1966); Biochim. Biophys. Acta,
     119:421 (1966).

50. S. Pelc, Nature, 207:597 (1965).

51. Yu. B. Rumer, Preprint: Sib. Otd. Akad. Nauk SSSR, Inst. Matematiki, Novo-
     sibirsk, 1965.

52. M. V. Vol'kenshtein and Yu. B. Rumer, Preprint TF-20: Sib. Otd. Akad. Nauk
     SSSR, Novosibirsk, 1965.

53. A. E. Braunshtein, Zh. Vses. Khim. Obshchestva im. D. I. Mendeleeva, 8(1):81
     (1963).

53a. A. E. Braunshtein, M. Ya. Karpeiskii, and R. M. Khomutov, in: Enzymes,
     Nauka, 1964, Chap. 9.

54. M. Dosher and F. Richards, Federation Proc., 22:419 (1963).

55. B. Harrap, D. Gratzer, and P. Doty, Ann. Rev. Biochim., 30:268 (1961).

56. K. Linderstrøm-Lang and J. Schellman, in: The Enzymes, Vol. I, Academic Press, New York, 1959, Chap. 10.
57. F. B. Straub and G. Szabolcsi, in: Molecular Biology: Problems and Prospects, Nauka, 1964, p. 182.
58. K. Okunuki, Advan. Enzymol., 23:29 (1961).
59. G. Szabolcsi and E. Biszku, Biochim. Biophys. Acta, 48:335 (1961).
60. J. Kirkwood and J. Shumaker, Proc. Natl. Acad. Sci. U. S., 38:863 (1952).
61. O. B. Ptitsyn and Yu. E. Eizner, Biofizika, 10:3 (1965).
62. A. Einstein, Ann. der Physik, 33:1275 (1965).
63. M. V. Vol'kenshtein, Molecular Optics, Gostekhizdat, 1951.
64. M. V. Vol'kenshtein, in: Molecular Biology (in Honor of the Seventieth Birthday of V. A. Engelhardt), Nauka, 1964, p. 172.
65. S. E. Bresler, Transactions of the Fifth International Biochemical Congress, Symposium 1, Izd. Akad. Nauk SSSR, 1962.
66. E. L. Smith, J. R. Kimmel, and A. Light, Transactions of the Fifth International Congress of Biochemistry, Vol. 4, Moscow, 1961, Macmillan, New York, 1963, p. 122.
67. E. Smith et al., J. Biol. Chem., 240:253 (1965).
68. J. Shevallier, Y. Jacquot-Armand, and J. Inn, Biochim. Biophys. Acta, 92:521 (1964).
69. S. E. Shnol', in: Molecular Biophysics, Nauka, 1965, p. 56.
70. N. Slater, Theory of Unimolecular Reactions, Cornell University Press, 1959.
71. S. W. Benson, The Foundations of Chemical Kinetics, McGraw-Hill, New York, 1960.
72. I. V. Obreimov, Zh. Éksperim. i Teor. Fiz., 19:396 (1949).
73. B. I. Stepanov, Luminescence of Complex Molecules, Izd. Akad. Nauk Belorussk. SSR, Minsk, 1955.
74. R. Lumry, in: The Enzymes, Vol. I, Academic Press, New York, 1959, Chap. 3.
75. F. Karush, J. Am. Chem. Soc., 72:2705 (1950).
76. C. H. Waddington, New Patterns in Genetics and Development, Columbia University Press, New York, 1962.
77. W. Boyd, Introduction to Immunochemical Specificity, Interscience (Wiley), New York, 1962.
78. F. M. Burnet, The Integrity of the Body, Harvard University Press, Cambridge, Mass., 1962.
79. L. Pauling, J. Am. Chem. Soc., 62:2643 (1940).
80. F. Vaslov and D. Doherty, J. Am. Chem. Soc., 75:928 (1953).
81. A. A. Balandin, The Multiplet Theory of Catalysis, Izd. Mosk. Gos. Univ., 1963.
82. A. A. Balandin, Dokl. Akad. Nauk SSSR, 114:1008 (1957).
83. A. A. Balandin, Biokhimiya, 23:475 (1958).
84. D. Koshland, in: The Enzymes, Vol. I, Academic Press, New York, 1959, Chap.7.
85. D. Koshland, Proc. Natl. Acad. Sci. U. S., 44:98 (1958).
86. D. Koshland, J. Yankeelov, and J. Thoma, Federation Proc., 21:1031 (1962).
87. D. Koshland, J. Theoret. Biol., 2:75 (1962).
88. I. Prigogine, The Molecular Theory of Solutions, North Holland Publishing Co., Amsterdam, 1963.

89. D. Koshland, Cold Spring Harbor Symp. Quant. Biol., 28 : 473 (1963).

90. M. Burr and D. Koshland, Proc. Natl. Acad. Sci. U. S., 52 : 1017 (1964).

91. D. Koshland, Y. Karkhanis, and H. Latam, J. Am. Chem. Soc., 86 : 1448 (1964).

92. S. Grisolia and B. Joyce, Biochim. Biophys. Res. Comm., 1 : 280 (1959).

93. G. Tomkins, K. Yielding, and J. Curban, Proc. Natl. Acad. Sci. U. S., 47 : 270
     (1961).

94. H. Schachman, Brookhaven Symp. Theoret. Biol., 17 : 91 (1964).

95. M. Citri and N. Garber, Biochim. Biophys. Res. Commun., 4 : 143 (1961).

96. G. Ning Ling, Biopolymers, Symposia, No. 1, 1964, p. 91.

97. W. P. Jencks, Ann. Rev. Biochem., 32 : 639 (1963).

98. B. Nagy and W. P. Jencks, Biochemistry, 1 : 987 (1962).

99. D. Koshland, in: Horizons in Biochemistry [Russian translation], Mir, 1964.
     [English edition: Horizons in Biochemistry, Albert Sgent-Györgyi Dedicatory
     Volume, edited by M. Kasha and B. Pullman, Academic Press, New York.]

100. J. Wooton and G. Hess, J. Am. Chem. Soc., 84 : 440 (1962).

101. B. Labouesse, B. Havsteen, and G. Hess, Proc. Natl. Acad. Sci. U. S., 48 : 2137
     (1962).

102. G. Fasman, K. Norland, and A. Pesce, Biopolymers, Symposia No. 1, 1964,
     p. 325.

103. J. Yankeelov and D. Koshland, J. Biol. Chem., 240 : 1593 (1965).

104. G. Stark, W. Stein, and S. Moore, J. Biol. Chem., 236 : 436 (1961).

105. D. Koshland, D. Strumeyer, and W. Ray, Brookhaven Symp. Theoret. Biol.,
     15 : 101 (1962).

106. L. Johnson and D. Phillips, Nature, 206 : 761 (1965).

107. D. C. Phillips, Abstracts of Papers Presented at the Seventh International Con-
     gress on Crystallography [Russian translation], Nauka, 1966, p. 153; D. C.
     Phillips, Sci. Am., November : 78 (1966).

108. M. Born, Optik, Springer, Berlin, 1965.

109. W. Kauzmann, Quantum Chemistry, Academic Press, New York, 1957.

110. H. Eyring, J. Walter, and G. E. Kimball, Quantum Chemistry, Interscience
     (Wiley), New York, 1944.

111. D. Caldwell and H. Eyring, Ann. Rev. Phys. Chem., 15 : 281 (1964).

112. G. Kirkwood, J. Chem. Phys., 5 : 479 (1937).

113. M. V. Vol'kenshtein, Dokl. Akad. Nauk SSSR, 71 : 447, 643 (1950).

114. A. McLachlan and M. Ball, Mol. Phys., 8 : 581 (1965).

115. V. M. Aslanyan and M. V. Vol'kenshtein, Opt. i Spektroskopiya, 7 : 208 (1959).

116. W. Kautzmann, J. Walter, and H. Eyring, Chem. Rev., 26 : 339 (1940).

117. E. Kondon, Usp. Fiz. Nauk, 19 : 380 (1938).

118. M. P. Kruchek, in: Optics and Spectroscopy, Vol. 2, Izd. Akad. Nauk SSSR,
     1963, p. 65; Opt. i Spektroskopiya, 17 : 545, 794 (1964).

119. M. V. Vol'kenshtein and I. O. Levitan, Zh. Strukt. Khim., 3 : 81, 87 (1962).

120. I. O. Levitan and M. V. Vol'kenshtein, in: Optics and Spectroscopy, Vol. 2,
     Izd. Akad. Nauk SSSR, 1963, p. 60.

121. C. Djerassi, Optical Rotatory Dispersion, McGraw-Hill, New York, 1960.

122. A. Moscowitz, Rev. Mod. Phys., 32 : 440 (1960); Advan. Chem. Phys., 4 : 67
     (1962).

123.   G. Holzwarth, W. Gratzer, and P. Doty, Biopolymers, Symposia No. 1, 1964, p. 389.

124.   W. Moffit and J. Yang, Proc. Natl. Acad. Sci. U. S., 42 : 596 (1956).

125.   P. Urnes and P. Doty, Advan. Protein Chem., 16 : 401 (1961).

126.   W. Leonard and J. Foster, J. Mol. Biol., 7 : 590 (1963).

127.   J. Yang, Proc. Natl. Acad. Sci. U. S., 53 : 438 (1965).

128.   W. Moffit, J. Chem. Phys., 25 : 467 (1956).

129.   A. S. Davydov, Theory of Light Absorption in Molecular Crystals, Izd. Akad. Nauk SSSR, Kiev, 1951; see also, A. S. Davydov, Theory of Molecular Excitons, McGraw-Hill, New York, 1962.

130.   J. Ham and P. Platt, J. Chem. Phys., 20 : 335 (1952).

131.   D. Peterson and W. Simpson, J. Am. Chem. Soc., 79 : 2375 (1957).

132.   I. Tinoco, A. Halpern, and W. Simpson, in: Polyamino Acids, Polypeptides, and Proteins, University of Wisconsin Press, 1962, p. 147.

133.   W. Moffitt, D. Fitts, and J. Kirkwood, Proc. Natl. Acad. Sci. U.S., 43:723 (1957).

133a. D. Fitts and J. Kirkwood, Proc. Natl. Acad. Sci. U. S., 43 : 1046 (1957).

134.   H. Murakami, J. Chem. Phys., 27 : 1231 (1957).

135.   I. Tinoco, J. Am. Chem. Soc., 82 : 4785 (1960).

136.   I. Tinoco, J. Chem. Phys., 33 : 1332 (1960).

137.   A. Rich and I. Tinoco, J. Am. Chem. Soc., 82 : 6409 (1960).

138.   H. DeVoe and I. Tinoco, J. Mol. Biol., 4 : 518 (1962).

139.   I. Tinoco, R. Woody, and D. Bradley, J. Chem. Phys., 38 : 1317 (1963).

140.   D. Bradley, I. Tinoco, and R. Woody, Biopolymers, 1 : 239 (1963).

141.   I. Tinoco and R. Woody, J. Chem. Phys., 32 : 461 (1960).

142.   I. Tinoco, in: Exciton Symposium, Radiation Research Soc. Meeting, Colorado Springs, 1962, p. 133.

143.   I. Tinoco, J. Am. Chem. Soc., 86 : 297 (1964).

144.   I. Tinoco, Advan. Chem. Phys., 4 : 113 (1962).

145.   I. Tinoco and C. Bush, Biopolymers, Symposia No. 1, 1964, p. 389.

146.   R. Nesbet, Mol. Phys., 7 : 221 (1963).

147.   H. DeVoe, J. Chem. Phys., 41 : 393 (1964); Biopolymers, Symposia, No. 1, 1964, p. 251.

148.   R. Harris, J. Chem. Phys., 43 : 959 (1965).

149.   J. Schellmann and A. Oriel, J. Chem. Phys., 37 : 2114 (1962).

150.   G. Holzwarth and P. Doty, J. Am. Chem. Soc., 87 : 218 (1965).

151.   E. Shechter and E. Blout, Proc. Natl. Acad. Sci. U. S., 51 : 794 (1964).

152.   E. Shechter, J. Carver, and E. Blout, Proc. Natl. Acad. Sci. U.S., 51:1029 (1964).

153.   K. Imahori, Biochim. Biophys. Acta, 37 : 336 (1960).

154.   A. Wada, M. Tsuboi, and E. Konishi, J. Phys. Chem., 65 : 1119 (1961).

155.   M. V. Vol'kenshtein and V. A. Zubkov, Biopolymers, 5 : 465 (1967).

156.   B. Pullman and A. Pullman, Quantum Biochemistry, Interscience (Wiley), New York, 1963.

157.   E. Blout and E. Stryer, J. Am. Chem. Soc., 83 : 1411 (1961).

158.   E. Blout, Tetrahedron, 13 : 123 (1961).

159.   I. A. Bolotina and M. V. Vol'kenshtein, in: Molecular Biophysics, Nauka, 1965, pp. 27, 36.

160. K. Konstantinavichus and M. V. Vol'kenshtein, Opt. i Spektroskopiya, 23: 80 (1967).

161. V. I. Permogorov, Author's abstract of Candidate's Dissertation, IVS Akad. Nauk SSSR, 1965.

162. V. I. Permogorov, Yu. S. Lazurkin, and S. E. Shmurak, Dokl. Akad. Nauk SSSR, 155: 1440 (1964).

163. M. K. Li, D. Ulmer, and B. Vallee, Proc. Natl. Acad. Sci. U. S., 47: 1155 (1961); Biochemistry, 1: 114 (1962).

164. Yu. N. Breusov, V. I. Ivanov, M. Ya. Karpeiskii, and Yu. V. Morozov, Biochim. Biophys. Acta, 92: 388 (1964).

165. Yu. M. Torchinskii and L. G. Koreneva, Biokhimiya, 28: 1087 (1963); 29: 780 (1964); Dokl. Akad. Nauk SSSR, 155(4) (1964); Biochim. Biophys. Acta, 79: 426 (1964); Yu. M. Torchinskii, Author's abstract of Doctoral Dissertation, AMN SSSR, 1965; L. G. Koreneva, Author's abstract of Candidate's Dissertation, MFTI, 1965.

166. A. E. Braunshtein, Yu. M. Torchinskii, and L. G. Koreneva, Abstracts of Papers Presented at the Sixth International Biochemical Congress, New York, 1964.

167. K. Aki, T. Takagi, T. Isemura, and T. Yamano, Biochim. Biophys. Acta, 122: 193 (1966).

168. A. Platt and C. Niemann, Proc. Natl. Acad. Sci. U. S., 50: 817 (1963).

169. B. Havsteen and G. Hess, J. Am. Chem. Soc., 84: 491 (1962).

170. H. Parker and R. Lumry, J. Am. Chem. Soc., 85: 483 (1963).

171. B. Labouesse, H. Oppenheimer, and G. Hess, Biochem. Biophys. Res. Commun., 14: 318 (1964).

172. I. A. Bolotina, M. V. Vol'kenshtein, and O. P. Chikalova-Luzina, Biokhimiya, 31: 241 (1966).

173. G. Schwert, A. Winer, P. Boyer, H. Lardy, and K. Myrback, in: The Enzymes, Vol. 7, Academic Press, New York, 1963.

174. G. Sabato and M. Ottesen, Biochemistry, 4(3): 422 (1965).

175. W. Novoa and G. Schwert, J. Biol. Chem., 236: 2150 (1961).

176. R. Alberty, J. Am. Chem. Soc., 75: 1925 (1953).

177. I. A. Bolotina, D. S. Markovich, P. Zavodskii, and M. V. Vol'kenshtein, Molekul. Biol., 1: 231 (1967); D. S. Markovich, P. Zavodskii, and M. V. Vol'kenshtein, Dokl. Akad. Nauk SSSR, 170: 204 (1966).

178. G. Schwert and A. Winer, in: The Enzymes, Vol. 7, Academic Press, New York, 1963.

179. S. Velick and C. Furfine, in: The Enzymes, Vol. 7, Academic Press, New York, 1963, p. 243.

180. E. Racker, in: Sulfur in Proteins, Academic Press, New York, 1959, p. 211.

181. P. Boyer and A. Schulz, in: Sulfur in Proteins, Academic Press, New York, 1959, p. 199.

182. I. A. Bolotina, M. V. Vol'kenshtein, P. Zavodskii, and D. S. Markovich, Biokhimiya, 31: 649 (1966).

183. I. A. Bolotina, D. S. Markovich, M. V. Vol'kenshtein, and P. Zavodskii, Biochim. Biophys. Acta, 132: 260 (1967).

184. I. Listovsky, C. Furfine, J. Betheil, and S. England, J. Biol. Chem., 240:4253 (1965).
185. D. S. Markovich, P. Zavodskii, and M. V. Vol'kenshtein, Biokhimiya, 31:873 (1966).
186. G. Sabato and N. Kaplan, J. Biol. Chem., 239:438 (1964).
187. B. Jirgensons, J. Am. Chem. Soc., 83:3161 (1961).
188. A. Rosenberg, H. Theorell, and T. Yonetani, Nature, 203:755 (1964).
189. T. Mansour, J. Biol. Chem., 238:2285 (1963).
190. D. S. Markovich, M. P. Mel'nikova, M. V. Vol'kenshtein, and S. A. Neifakh, Biokhimiya, 31:1225 (1966).
191. J. L. Webb, Enzyme and Metabolic Inhibitors, Academic Press, New York, 1963.
192. M. Dixon and E. C. Webb, Enzymes, Academic Press, New York, 1964.
193. L. Michaelis and H. Davidsohn, Biochem. Z., 35:386 (1911).
194. J. Kirkwood, Disc. Faraday Soc., 20:78 (1955).
195. J. Kirkwood, Symposium on the Mechanism of Enzyme Action, Johns Hopkins University Press, Baltimore, 1954.
196. S. Timasheff, Biopolymers, 4:107 (1966).
197. W. Sheider, Biophys. J., 5:617 (1965).
198. M. V. Vol'kenshtein and S. N. Fishman, Biofizika, 11:956 (1966).
199. M. V. Vol'kenshtein and S. N. Fishman, Biofizika, 12:14 (1967).
200. B. Zimm and S. Rice, Mol. Phys., 3:391 (1960).
201. J. Applequist and P. Doty, in: Polyamino Acids, Polypeptides, and Proteins, edited by M. Stahmann, University of Wisconsin Press, 1962.
202. S. Lowey, J. Biol. Chem., 240:2421 (1965).
203. P. Doty, K. Imahori, and E. Klemperer, Proc. Natl. Acad. Sci. U. S., 44:424 (1958).
204. E. Blout and M. Idelson, J. Am. Chem. Soc., 80:4909 (1958).
205. M. Joly, A Physicochemical Approach to the Denaturation of Proteins, Academic Press, New York, 1965.
206. M. V. Vol'kenshtein and O. B. Ptitsyn, Zh. Tekh. Fiz., 25:649 (1955).
207. M. V. Vol'kenshtein, Biofizika, 6:257 (1961); Biophys. J., 2:189 (1962).
208. P. V. Afanas'ev, Biokhimiya, 14:424 (1949).
209. A. G. Pasynskii, Biophysical Chemistry, Vysshaya Shkola, Moscow, 1965.
210. M. V. Vol'kenshtein, in: Molecular Biophysics, Nauka, 1965, p. 5.
211. V. A. Yakovlev, Kinetics of Enzymatic Catalysis, Nauka, 1965.
212. V. A. Yakovlev, in: Enzymes, Nauka, 1965, Chap. 3.
213. A. Tiselius and I. B. Ertesson-Quensel, Biochim. J., 33:1752 (1930).
214. K. Linderstrøm-Lang, Cold Spring Harbor Symp. Quant. Biol., 14:117 (1950).
215. M. V. Vol'kenshtein, Dokl. Akad. Nauk SSSR, 100:468 (1965).
216. F. Jacob and J. Monod, Elements in Regulatory Circuits in Bacteria, Paper presented to a UNESCO Symposium on Biological Organization, September 1962.
217. F. Jacob and J. Monod, Transactions of the Fifth International Congress of Biochemistry, Vol. 1, Moscow, 1961, Macmillan, New York, 1963, p. 132.
218. F. Jacob, S. Brenner, and F. Cuzin, Cold Spring Harbor Symp. Quant. Biol., 28:329 (1963).
219. H. Umbarger, Science, 123:848 (1956); 145:674 (1964).

220. J. Gerhart and A. Pardee, Cold Spring Harbor Symp. Quant. Biol., 28:491 (1963).

221. M. V. Vol'kenshtein, in: Molecular Biophysics, Nauka, 1965, p. 16.

222. J. Monod, J.-P. Changeux, and F. Jacob, J. Mol. Biol., 6:306 (1963).

223. J.-P. Changeux, Cold Spring Harbor Symp. Quant. Biol., 28:313 (1963); J. Mol. Biol., 4:220 (1962).

224. J. Monod, J. Wyman, and J.-P. Changeux, J. Mol. Biol., 12:88 (1965).

225. F. Oosawa, S. Asabura, and T. Ooi, "Physical chemistry of muscle protein, actin." Progr. Theor. Phys., No. 17 (1961).

226. M. V. Vol'kenshtein and B. N. Gol'dshtein, Biochim. Biophys. Acta, 115:478 (1966).

227. C. Doy, Biochim. Biophys. Acta, 118:173 (1966).

228. D. Atkinson, Science, 150:851 (1965).

229. K. A. Kafiani, in: Biological Aspects of Cybernetics, Izd. Akad. Nauk SSSR, 1962, p. 210; in: Advances in Biological Chemistry, Vol. 5, Izd. Akad. Nauk SSSR, 1963, p. 100; in: Enzymes, Nauka, 1964, Chap. 10.

230. L. Pauling, Proc. Natl. Acad. Sci. U. S., 21:186 (1935).

231. A. Rossi-Fanelli, E. Antonini, and A. Caputo, Advan. Protein Chem., 19:73 (1964),

232. J. Wyman, Advan. Protein Chem., 4:407 (1948); 19:223 (1964).

233. J. Wyman, Cold Spring Harbor Symp. Quant. Biol., 28:483 (1963).

234. R. Benesch and Ruth Benesch, J. Mol. Biol., 6:498 (1963).

235. O. Gibson, Biochim. J., 71:193 (1959).

236. E. Antonini, J. Wyman, R. Moretti, and A. Rossi-Fanelli, Biochim. Biophys. Acta, 71:124 (1965).

237. Ruth Benesch and R. Benesch, Biochemistry, 1:735 (1962).

238. J. Wyman and D. Allen, J. Polymer Sci., 7:499 (1951).

239. R. Benesch and Ruth Benesch, J. Biol. Chem., 236:405 (1961).

240. A. A. Grinberg, Introduction to the Chemistry of Complex Compounds, Khimiya, 1966.

241. A. Rossi-Fanelli, E. Antonini, and A. Caputo, J. Biol. Chem., 236:391, 397 (1961).

242. R. Benesch, Biochemistry, 3:1132 (1964).

243. Ruth Benesch, R. Benesch, and G. Macduff, Proc. Natl. Acad. Sci. U. S., 54:535 (1965).

244. L. A. Blyumenfel'd, Hemoglobin and Reversible Oxygen Attachment, Sovetskaya Nauka, 1957.

245. J. Kendrew, Sci. Am., 205:96 (1961).

246. H. Muirhead and M. Perutz, Nature, 199:633 (1963).

247. R. Briehl, Federation Proc., 21:72 (1962).

248. M. V. Vol'kenshtein and A. K. Shemelin, Biofizika, 11:773 (1966).

249. M. V. Vol'kenshtein, Yu. A. Sharonov, and A. K. Shemelin, Molekul. Biol., 1:467 (1967).

250. M. V. Vol'kenshtein and A. K. Shemelin, Biokhimiya, 30:148 (1965).

251. V. A. Engelhardt and M. N. Lyubimova, Nature, 144:668 (1939); Biokhimiya, 4:716 (1939); 7:205 (1942).

252. V. A. Engelhardt, Usp. Sovrem. Biol., 14:177 (1941).

253. V. A. Engelhardt, Izv. Akad. Nauk SSSR, Ser. Biol., No.2, p. 182 (1945).

254. R. Davies, Nature, 199 : 1068 (1963).

255. Yu. Tonomura, T. Kanasawa, and K. Sekia, in: Molecular Biology (in Honor
      of the Seventieth Birthday of V. A. Engelhardt), Nauka, 1964, p. 213.

256. M. V. Vol'kenshtein, Dokl. Akad. Nauk SSSR, 146 : 1426 (1962).

257. V. I. Vorob'ev and L. V. Kukhareva, Dokl. Akad. Nauk SSSR, 165 : 435 (1965).

258. B. F. Poglazov, Structure and Function of Contractile Proteins, Nauka, 1965.

259. N. Silvester and M. Holwill, Nature, 205 : 665 (1965).

260. A. Lehninger, Physiol. Rev., 42 : 467 (1962).

261. T. V. Venkstern and V. A. Engelhardt, Dokl. Akad. Nauk SSSR, 102 : 133 (1955).

262. S. N. Neifakh, T. V. Kazakova, M. P. Mel'nikova, and V. S. Turovskii, Dokl.
      Akad. Nauk SSSR, 138 :227 (1961).

263. S. A. Neifakh and T. V. Kazakova, Nature, 197 : 1106 (1963).

264. S. A. Neifakh, in: Molecular Biology (in Honor of the Seventieth Birthday of
      V. A. Engelhardt), Nauka, 1964, p. 273.

265. S. A. Neifakh and V. S. Repin, Biochim. Biophys. Res. Commun., 14 : 86 (1963).

266. D. Green, in: Molecular Biology (in Honor of the Seventieth Birthday of V. A.
      Engelhardt), Nauka, 1964, p. 260.

267. H. Noll, T. Staehelin, and F. Wettstein, Nature, 198 : 632 (1963).

268. M. V. Vol'kenshtein and S. N. Fishman, Dokl. Akad. Nauk SSSR, 160 : 1407
      (1965).

269. B. Katz, Rev. Mod. Phys., 31 :269 (1959).

270. L. Opit and J. Charnock, Nature, 208 : 471 (1965).

271. E. King and C. Altman, J. Phys. Chem., 60 : 1375 (1956).

272. E. King, J. Phys. Chem., 60 : 1378 (1956).

273. W. Cleland, Biochim. Biophys. Acta, 67 : 104, 173 (1963).

274. M. V. Vol'kenshtein and B. N. Gol'dshtein, Biochim. Biophys. Acta, 115 : 471
      (1966).

275. S. J. Mason and H. J. Zimmerman, Electronic Circuit Theory, Interscience
      (Wiley), New York, 1960.

276. L. Robichaud, M. Buaver, and J. Robert, Directional Graphs [Russian translation],
      Energiya, 1964; Signal Flow Graphs and Other Applications, Prentice-Hall (1962).

277. S. Mason, Proc. IRE, 41 : 1144 (1953); 44 : 920 (1956).

278. P. A. Ionkin and A. A. Sokolov, Elektrichestvo, 5 : 67 (1964).

279. P. A. Ionkin, Elektrichestvo, 8 : 26 (1964).

280. M. V. Vol'kenshtein and B. N. Gol'dshtein, Dokl. Akad. Nauk SSSR, 170 : 963
      (1966).

281. J. Wong and C. Hanes, Can. J. Biochim. Physiol., 40 : 763 (1962).

282. J. Wong and C. Hanes, Nature, 203 : 492 (1964).

283. J. Wong, J. Am. Chem. Soc., 87 : 1788 (1965).

284. M. V. Vol'kenshtein, B. N. Gol'dshtein, and V. E. Stefanov, Molekul. Biol.,
      1 : 52 (1967).

285. B. Chance, D. Garfinkel, J. Higgins, and B. Hess, J. Biol. Chem., 235 :2426
      (1960).

286. V. A. Ditkin and A. P. Prudnikov, Operational Computation, Vysshaya Shkola,
      1966; Handbook of Operational Computation, Vysshaya Shkola, 1965.

287. E. Strickland and E. Ackerman, Nature, 209: 405 (1966).
288. J. Christiansen, Advan. Catalysis, 5: 311 (1953).
289. M. I. Temkin, Dokl. Akad. Nauk SSSR, 152: 156 (1963).
290. M. I. Temkin, Dokl. Akad. Nauk SSSR, 165: 615 (1965).
291. M. Faraday, Experimental Researches in Electricity, Vol. 3, London, 1839-55.
292. D. Macaluso and O. Corbino, Nuovo Cimento, 8: 257 (1898); 9: 381 (1899).
293. R. W. Wood, Physical Optics, Macmillan, New York, 1934.
294. R. Serber, Phys. Rev., 41: 489 (1932).
295. I. Tobias and W. Kauzmann, J. Chem. Phys., 35: 538 (1961).
296. M. Groenevege, Mol. Phys., 5: 541 (1962).
297. C. J. Ballhausen, Introduction to Ligand Field Theory, McGraw-Hill, New York, 1962.
298. Ya. G. Dorfman, Diamagnetism and the Chemical Bond, Fizmatgiz, 1961.
299. R. Becker, The Classical Theory of Electricity and Magnetism, Hafner, New York, 1950.
300. W. Eberhardt, Wu-Chien Cheng, and H. Renner, J. Mol. Spectry., 3: 664 (1959).
301. W. Eberhardt and H. Renner, J. Mol. Spectry., 6: 483 (1961).
302. W. Eberhardt and B. Snowden, J. Mol. Spectry., 18: 372 (1965).
303. D. Hameka, J. Chem. Phys., 36: 2540 (1962).
304. V. Shashoua, J. Am. Chem. Soc., 82: 5505 (1960); 86: 2109 (1964).
305. V. Shashoua, Nature, 203: 972 (1964); Biochemistry, 3: 1719 (1964); J. Am. Chem. Soc., 87: 4044 (1965).
306. M. V. Vol'kenshtein, Yu. A. Sharonov, and A. K. Shemelin, Nature, 209: 709 (1966).
307. M. A. El'yahsevich, Spectra of the Rare Earths, Gostekhizdat, 1953, p. 115.
308. L. D. Landau and E. M. Lifshits, Quantum Mechanics, Nauka, 1964.
309. D. Hameka and L. Oosterhoff, Mol. Phys., 1: 358 (1958).
310. I. Tinoco and C. Bush, Biopolymers, Symposia, 1: 209 (1964).
311. P. Stephens, W. Suetaak, and P. Schatz, J. Chem. Phys., 44: 4592 (1966).
312. P. Pershan, M. Gouterman, and R. Fulton, Mol. Phys., 10: 397 (1966).
313. B. P. Atanasov, M. V. Vol'kenshtein, Yu. A. Sharonov, and A. K. Shemelin, Molekul. Biol., 1: 477 (1967).
314. D. Urry and H. Eyring, Proc. Natl. Acad. Sci. U. S., 49: 253 (1963); J. Theoret. Biol., 8: 198, 214 (1965).
315. V. P. Atanasov, International Symposium on Comparative Hemoglobin Structure, Thessaloniki, 1966, p. 13.
316. R. Benesch, Ruth Benesch, and I. Tyuma, Proc. Natl. Acad. Sci. U. S., 56: 1268 (1966).
317. C. Bigelow, J. Theoret. Biol., 16: 187 (1967).
318. F. H. C. Crick, J. Mol. Biol., 19: 548 (1966).
319. D. S. Chernavskii, Yu. I. Khurgin, and S. E. Shnol', Molekul. Biol., 1: 419 (1967).
320. M. V. Vol'kenshtein, Dokl. Akad. Nauk SSSR (in press).
321. J. P. Changeux et al., Proc. Natl. Acad. Sci. U. S., 57: 334 (1967).
322. M. V. Vol'kenshtein, L. Govshovichus, Yu. A. Sharonov, and A. K. Shemelin, Molekul. Biol., 1(6) (1967).
323. A. Buckingham and P. Stephens, Ann. Rev. Phys. Chem., 17: 399 (1967).

INDEX